Energy for the Future

Technology in the Third Millennium

1. **The Treatment and Handling of Wastes**
   Edited by A.D. Bradshaw, Sir Richard Southwood and Sir Frederick Warner
2. **A Global Strategy for Housing in the Third Millennium**
   Edited by W.A. Allen, R.G. Courtney, E. Happold and Sir Alan Muir Wood
3. **Communications after AD 2000**
   Edited by D.E.N. Davies, C. Hilsum and A.W. Rudge
4. **Health Policy and Technological Innovation**
   Edited by J. Newson-Davis and Sir David Weatherall
5. **Passenger Transport after 2000 AD**
   Edited by G.B.R. Feilden, A.H. Wickens and A.R. Yates
6. **Energy for the Future**
   Edited by Sir Denis Rooke, Ian Fells and John Horlock

Energy for the Future

Edited by
Sir Denis Rooke,
Ian Fells and
John Horlock

Published by E & FN Spon for The Royal Society

The Royal Society

E & FN SPON
An Imprint of Chapman & Hall
London · Glasgow · Weinheim · New York · Tokyo · Melbourne · Madras

Published by E & FN Spon, an imprint of Chapman & Hall, 2–6 Boundary Row, London SE1 8HN, UK

Chapman & Hall, 2–6 Boundary Row, London SE1 8HN, UK

Blackie Academic & Professional, Wester Cleddens Road, Bishopbriggs, Glasgow G64 2NZ, UK

Chapman & Hall GmbH, Pappelallee 3, 69469 Weinheim, Germany

Chapman & Hall USA, One Penn Plaza, 41st Floor, New York NY 10119, USA

Chapman & Hall Japan, ITT-Japan, Kyowa Building, 3F, 2–2–1 Hirakawacho, Chiyoda-ku, Tokyo 102, Japan

Chapman & Hall Australia, Thomas Nelson Australia, 102 Dodds Street, South Melbourne, Victoria 3205, Australia

Chapman & Hall India, R. Seshadri, 32 Second Main Road, CIT East, Madras 600 035, India

First edition 1995

Typeset in 10/12 pt Times by Colset Pte Ltd, Singapore

Printed in Great Britain at the University Press, Cambridge

ISBN 0 419 20050 9

A catalogue record for this book is available from the British Library

Library of Congress Catalog Card Number: 94–74709

Printed on permanent acid-free text paper, manufactured in accordance with ANSI/NISO Z39.48–1992 and ANSI/NISO Z39.48–1984 (Permanence of Paper).

Contents

Contributors

K.J. Brown
Energy Technology Support Unit
8156 Harwell Laboratory
Harwell
OX11 0RA, UK

D.A. Davis
Metchley Cottage
Stony Lane
Little Kingshill
Gt. Missenden
Bucks
HP16 0DS, UK

R. Hawley
Chief Executive
Nuclear Electric plc
Barnett Way
Barnwood
Glos
GL4 7RS
UK

Professor J.H. Horlock
Whittle Laboratory
Madingley Road
Cambridge
CB3 0DY, UK

J. Masters
Programme Controller, Industrial
and Commercial Utilization
R&D Division, British Gas plc
Midland Research Station
Wharf Lane
Solihull B91 2JW
UK

K.K. Pillai
ABB Carbon AB
Pressurized Fluidized Bed Plants
S-612 20 Finspong
Sweden

Sir Denis Rooke
1 Great Cumberland Place
Marble Arch
London
W1P 9LN
UK

R. Carle
Directeur General Adjoint
Electricité de France
32 Rue de Monceau
75008 Paris
France

Professor I. Fells
Dept of Chemical and Process Engineering
University of Newcastle-upon-Tyne
Merz Court
Newcastle-upon-Tyne
NE1 7RU
UK

Professor P.J. Hills
Department of Transport Engineering
University of Newcastle-upon-Tyne
Newcastle-upon-Tyne
NE1 7RU
UK

E. Lees
Chief Executive
Energy Saving Trust
11/12 Buckingham Gate
London
SW1E 6LB
UK

Professor D. Pearce
Department of Environmental Economics
University College London
Gower Street
London
WC1E 6BT
UK

G. Preston
Vice-President, Generation
Electric Power Research Institute
3412 Hillview Avenue
PO Box 10412, Palo Alto
CA 94305
USA

Preface

Sir Denis Rooke CBE, FEng, FRS

The Royal Society has initiated a series of meetings to examine the ways in which technology may contribute to meeting human needs and expectations in the 21st century. One of the most basic of these needs, indeed akin to food, is energy and, in recognition of the wide scope of the subject, the Royal Academy of Engineering joined with the Royal Society in sponsoring a two-day meeting on the theme on the 16 and 17 March 1994.

The relentless growth of world population, most of it occurring within developing countries, and the legitimate push of those nations for improvements in their economic circumstances and towards industrial development presupposes an ever-increasing demand for energy. The wealthy nations consume the overwhelming proportion of currently available energy, not necessarily with conspicuous efficiency, but the present regional imbalances cannot continue indefinitely and the efficiency in use needs to be improved continuously.

No single right solution can be propounded since the optimum solution for each country/region needs to be found in the context of its particular physical, economic and political parameters. Neither can it be foreseen that purely technological advances will suffice to meet the need. Engineers and technologists, while having an important part to play will have to co-operate with specialists from many other disciplines if the social and environmental constraints are to be met.

As the era of cheap oil and gas comes to an end, in the face of the increasing demand all primary energy sources capable of economic exploitation will have to be brought into use, even including nuclear energy, unless fundamentally different lifestyles requiring much less energy to sustain them are to be adopted, which seems a rather remote possibility. This will in turn entail a heavy demand for development capital and satisfying that need will require political agreement on an international scale.

Having regard to the lead times for technological development and to the substantial non-technological barriers to be overcome, it will be impracticable to rely solely on the mechanisms of the market to bring about the required transformation in the global energy economy. The remaining years of the 20th century should be used for formulating governmental policies which, while taking cognizance of the market place, are deliberately based upon longer term considerations to aid the transition to robust patterns of energy supply and efficient usage world-wide.

The energy scenario is vast and the chapters that follow address just some aspects of the problems to be studied and overcome as we move into the third millennium.

Sir Denis Rooke CBE, FEng, FRS

1

Growth of energy demand and resources

D.A. Davis

1.1 INTRODUCTION

Energy and food are similar in their basic characteristics as world commodities. Each represents about 10% of total world trade across borders. In neither is there a shortage in absolute terms, nor is there likely to be in the foreseeable future. But in both there are grave deficiencies in distribution and availability to large tranches of the population.

In the longer term, growth in energy demand will be driven by increase in population, of which over 90% will take place in the developing countries. Actual growth rate will depend upon the extent to which populations are able to improve efficiency of end use and of energy production and conversion. It will also depend upon energy policies, not least in regard to energy pricing in the light of environmental and financial concerns and of depletion of traditional fossil-based resources.

This chapter examines possible alternative developments in energy demand and supply and their impact upon wider environmental and social issues – not least upon the energy poverty experienced by a majority of the world's population today. The purpose of this chapter is to provide an outline of the size and nature of the global need for energy through the next 30 years, extending this rather speculatively to the end of the 21st century, and to set this against current understanding of the energy likely to be available to us using technology and potential developments known to us today. However, that is only part of the task and in the author's view not the most important part.

The preface to this meeting implies that 'the quality of life depends to a great extent on the availability of energy refers, absolutely rightly, to the problems of demand growth, to depletion of oil and gas reserves and to environmental constraints. Two vital points hidden within these statements need to be highlighted. First, energy is a means to an end. It has no virtue other than facilitating the provision of things essential – or pleasant – to our well-being: heat, light, food, clothing, shelter, transport, medical care, fun. Energy is not an end in itself, any more than is money or the environment. These are to be valued or ignored as we will: the value they have is that which we ascribe to the things they are able to provide. So in thinking about energy and technology we need to include not only the technology of energy extraction, transformation and application but the technology (and, incidently, all the social/political implications) of the 'added value to life' that energy is seen to provide. In simple terms, there are more ways of getting fresh milk than keeping a refrigerator. Technology has to be applied to meeting people's need for the services energy provides as well as to the exploitation of energy itself.

Second, the global problem with energy, now and into the longer term, is the same as the global problem with food. They run parallel in their global characteristics. There is no absolute shortage of either nor is there likely to be. But there are parts of the world – and it is not necessary to remind you of some prime examples – where there is little food and little energy available to starving and deprived people. So while it is necessary to start with the traditional macro energy/demand balance we need to be much more concerned with energy distribution than about 'resources' as such. I shall therefore devote as many words to the latter as to the former.

1.2 ENERGY RESOURCES

1.2.1 Fossil fuels

The uncertainties associated with what fossil energy resources could be available for exploitation are quite different from the uncertainties associated with non-fossil sources. In the case of fossil fuels it is important to distinguish between proved recoverable reserves of fossil fuels and ultimately recoverable resources, which are defined as the potential output assuming high but not prohibitive prices and no insuperable barriers to exploitation. The uncertainties are largely technical – whether or not we can reasonably predict the costs of extraction and processing – and sociopolitical.

Proved recoverable fossil fuel reserves (Table 1.1) amount to about 850 billion tonnes of oil equivalent (Gtoe) and suggest reserves to production (R/P) values for coal, oil and gas of 197, 40 and 56 years respectively. The figures for ultimately recoverable resources (Table 1.2) summate to nearly 4500 Gtoe and would extend the resources to current production ratios by about 2 to 5 times. In practice R/P ratios have not fallen significantly over

Table 1.1 Proved recoverable fossil fuel reserves

<table>
<tr><th></th><th>Estimate of cumulative production to 1990 (Gtoe)</th><th>Estimate of proven reserves in 1990 (Gtoe)</th><th>Estimate of 1990 reserves to production ratio (years)</th></tr>
<tr><td>Coal (excluding lignite)</td><td rowspan="2">130</td><td>474</td><td>197</td></tr>
<tr><td>Lignite</td><td>110</td><td>293</td></tr>
<tr><td>Oil</td><td>86</td><td>137</td><td>40</td></tr>
<tr><td>Natural gas</td><td>40</td><td>108</td><td>56</td></tr>
</table>

Source: WEC Survey of Energy Sources.

Table 1.2 Ultimately recoverable fossil fuel reserves

	Gtoe	*%*
Coal and lignite	3400	76
Conventional oil	200	5
Unconventional oils:		
Heavy crude	75	2
Natural bitumen	70	2
Oil shale	450	10
Natural gas	220	5
Total (approx.)	4400	100

Source: WEC Survey of Energy Resources *et al.*

the last decade in spite of rapidly increasing production, simply because new finds have outpaced demand. There can be no precision about the future course of R/P ratios but leaving aside political influences the size of the figures suggests that fossil fuels are unlikely to escalate significantly in price because of reducing R/P ratios for some decades although there is likely to be a drift upwards. However, fossil fuels are finite and coal is the only fossil fuel likely to be available in substantial quantities much beyond the middle of the 21st century.

1.2.2 Nuclear power

Today there are about 420 commercial nuclear power plants in operation world-wide with a total generating capacity of about 340 GW. These contribute about 0.4 Gtoe or about 4.5% to global energy provision. Plants under construction are expected to add another 7 GW or so by the year

2000. The potential contribution of nuclear power to energy need over the next 30 years or so is presently limited by sociopolitical attitudes and the major uncertainty about nuclear power in the medium term is how these attitudes might change. If it is assumed that all the plant currently in service will be time expired by 2020 and a maximum commissioning rate over the next 30 years or so of about 30 GW per year can be achieved, then the nuclear contribution by 2020 would rise to about 1.0 Gtoe. But even that must depend on whether or not society rejects or accepts its widespread exploitation.

I need hardly emphasize here the potential of nuclear energy. The present rate of uranium use is approaching 58 000 tonnes per year. Known resources recoverable at less than US$80/kg are equivalent to around 40 years at current requirements, assuming they could all be exploited and also assuming a once through cycle. This extends to around 60 years if uranium resources recoverable at less than US$130/kg are included. Reprocessing and recycling of uranium and plutonium in thermal reactors could extend this time period further, but the fact remains that a substantial escalation in thermal reactor commissioning would give real substance to the need to bring back on course the commercial development of fast reactor technology for exploitation in the third or fourth decades of the 21st century or even earlier. Indeed, having been involved in and campaigned hard and long in the United Kingdom for the European Fast Reactor Project, and lost, the author is much in favour of such an expansion of thermal reactor exploitation which would be likely to provide the clear cut economic justification for continued fast reactor development on a global scale that has thus far eluded us. The potential of the uranium reserves, including assuming eventual use in fast reactors, is enormous, and the figures indicate equivalent energy reserves nearly twice the estimate for all fossil fuels. (Remy Carle deals with nuclear power in greater depth in Chapter 3.)

1.2.3 Renewables

The major uncertainties in the development of renewable energy forms also include sociopolitical factors – to what extent substantial policy support for their exploitation might be forthcoming – and environmental factors since some forms do raise environmental concerns. Of the 18% of the world's primary energy supply which came from all renewable sources in 1990 over 90% came from traditional biomass and large hydro. The future of renewable energy must lie in the 'new renewables' – new biomass, solar, geothermal and wind. These contribute less than 2% of primary energy supply at the present time.

The Renewable Energy Resources Committee of the World Energy Council has provided a range of possibilities to the year 2020 which apart from the factors mentioned above are also demand dependent. They suggest a possible range of 539–1345 Mtoe, which, given demand uncertainty (see below), would represent, say, 3–12% of demand. Table 1.3 summarizes their figures. Dr Kevin Brown, in Chapter 7, deals in greater depth with these matters. What is important in the context of this overview is the order of magnitude of the figures and to recognize their dependence on political attitudes.

1.3 ENERGY DEMAND

Energy demand growth (or decline) into the longer term will depend on a small number of simple, interlinked elements whose behaviour over time depends upon highly complex factors. The primary elements are:

1. population growth and its distribution;
2. energy per capita consumption;
3. economic growth;
4. energy intensity.

Table 1.3(a) Estimated contribution of 'new renewable' energy resources

	1990	*2020 (minimum Mtoe)*	*2020 (maximum Mtoe)*
Modern biomass	121	243	561
Solar	12	109	355
Wind	1	85	215
Geothermal	12	40	91
Small hydro	18	48	69
Ocean/tidal	0	14	54
Total	164	539	1345

Source: WEC Renewable Energy Resources Committee.

Table 1.3(b) Estimated contribution of 'other renewable' energy resources

	1990	*2020 Minimum mtoe*	*2020 Maximum mtoe*
Traditional biomass	930	1320	1060
Large hydro	465	980	875
Total	1395	2300	1935

Source: WEC Renewable Energy Resources Committee.

Given these, qualitative judgements need to be applied to provide justifiable patterns of demand and supply, such judgements concern, for example:

1. the desire for the services which energy provides;
2. technological innovation and transfer;
3. the availability of finance;
4. institutional reform;
5. societal attitudes and their influence on fuel mix, not least towards environmental protection.

1.3.1 Population growth

Annual growth rate in world population has risen steadily during the 20th century - and well before. The world population was just over 1. 6 billion in 1900 and had risen to 3.9 billion by 1975 and to 5.3 billion by 1990. The United Nations' central estimate is 1.4% p.a. into the 21st century: applied through to 2100 this would result in the population rising as shown in Table 1.4. United Nations quotes a range from 10.5 to 20 billion in 2100. In the shorter term a more important element is the geopolitical distribution of population growth. In 1990 the developing countries accounted for 76% of the population but over 90% of growth in the next few decades is likely to be in those countries.

1.3.2 Energy per capita consumption

There is a wide disparity in the distribution of energy available and used across the globe. Some 70% of the world's population exists at a per capita energy consumption which is barely above subsistence level - less than 1.2 tonnes oil equivalent (Table 1.5). The world average figure is 1.65 toe: in the US the figure is 7.9 toe per capita and in Western Europe 3.2 toe per capita. While of course these statistics depend to some extent on the efficiency of energy use the, message is clear: too many people have inadequate access to the energy supplies they need and 50% of the world's population have no access to commercial energy at all. Over 50 countries have experienced a fall in energy per capita consumption over the past 20–30 years. Apart from the distance travelled and time required to collect traditional fuels and the impact on health in their use, the low efficiency of use of non-commercial energy in poorer countries means that the energy

Table 1.4 World population (billions)

Year	*1990*	*2020*	*2050*	*2100*
Population	5.3	8.1	10.1	12

Table 1.5 Primary energy per capita (tonnes of oil equivalent per capita)

	1960	*1970*	*1980*	*1990*
North America	5.8	7.8	7.9	7.8
Western Europe	1.8	2.6	3.0	3.2
Commonwealth of Independent States	2.1	3.0	4.1	5.0
Central and Eastern Europe (CEE)	1.7	2.6	3.5	2.9
Latin America	0.7	0.9	1.2	1.3
Middle East	0.3	0.5	0.8	1.2
Sub-Saharan Africa	0.4	0.5	0.6	0.5
Pacific	0.5	0.6	0.8	1.0
South Asia	0.7	0.2	0.3	0.4
World	1.1	1.4	1.6	1.7

Source: WEC.

available at the point of use is very much lower than the per capita figures imply.

While there is and will continue to be growing pressure to increase per capita consumption in the developing countries, the implication of population growth and the low per capita energy use base in the developing countries is that without massive investment in an energy infrastructure in the developing areas of the world it is hard to see how an increase in energy supply can do any more than keep up with population growth and modest economic growth, let alone make a substantial contribution to improving per capita energy consumption. The rate of increase in the availability of energy per capita over the next 30 years is likely in some instances to be minimal.

1.3.3 Economic growth

Historic growth rates since 1960 are shown in Table 1.6. For the purposes of demand speculation it would seem reasonable – but always arguable – that, while there remains room for great volatility on a per annum per country basis in CEE/CIS and particularly in the developing countries over

Table 1.6 Historic growth rates (GDP % growth per annum)

	60–65	*65–70*	*70–75*	*75–80*	*80–85*	*85–90*
OECD	5.3	4.5	3.0	3.4	2.5	3.4
CEE/CIS	5.0	5.1	4.1	2.3	2.4	0.8
DCs	4.1	6.5	6.2	5.4	3.6	4.1
World	4.9	5.1	4.0	3.9	2.8	3.4

the next few decades, growth rates in OECD countries and in the CEE/CIS will stabilize within a range 2–3% per annum and in the developing countries will lie within a range 4–6% per annum.

1.3.4 Energy intensity

Energy intensity – the incremental change in energy consumed per incremental change in GNP – is influenced by factors, such as change in the structure of economic activities and in capital stock, that are not necessarily a direct reflection of the efficiency of energy use. However, with growing awareness and concern about our use of energy, not least in relation to the environment, policy measures directed towards increasing energy efficiency and conservation – economic incentives or disincentives, tough regulatory regimes – are likely to continue the historic trend in reducing energy intensity. The historic pattern is as shown in Table 1.7. Energy intensity is described further below.

1.3.5 Energy demand projections

In order sensibly to accommodate these elements and the possible range of numerical change that might be attributed to them, four 'cases' are presented for the period up to the year 2020, three of them being extended in less detail through to the year 2100. They are:

1. a 'reference' case;
2. a 'high growth' case;
3. a 'modified reference' case as a subset of the reference;
4. an 'ecologically driven' case;

Table 1.8 shows the main elements of these four cases for the period up to 2020. All four cases assume global population growth to be in line with the mean UN projection at 1.4%. Average economic growth is taken as 2.4% p.a. in all cases for OECD and for Eastern Europe and the CIS, but two possibilities for the developing countries are assumed: 5.6% p.a. as a high case and 4.6% p.a. as a moderate case. Differing assumptions for

Table 1.7 Historic growth rates of change in energy intensity (% p.a.)

	60–65	*65–70*	*70–75*	*75–80*	*80–85*	*85–90*
OECD	0.2	0.1	−1.3	−1.5	−2.1	−1.6
CEE/CIS	0.1	0.0	0.3	0.9	0.2	−0.5
DCs	−0.6	−2.3	−0.8	0.4	0.2	0.1
World	0.0	−0.5	−0.9	−0.9	−0.7	−1.1

Table 1.8 The four energy cases

Case	*A*	*B1*	*B*	*C*
Name	*High growth*	*Modified reference*	*Reference*	*Ecologically driven*
Economic growth (% p.a.)	High	Moderate	Moderate	Moderate
OECD	2.4	2.4	2.4	2.4
CEE/CIS	2.4	2.4	2.4	2.4
DCs	5.6	4.6	24.6	4.6
World	3.8	3.3	3.3	3.3
Energy intensity reduction (% p.a.)	High	Moderate	High	Very High
OECD	−1.8	−1.9	−1.9	−2.8
CEE/CIS	−1.7	−1.2	−2.1	−2.7
DCs	−1.3	−0.8	−1.7	−2.1
World	−1.6	−1.3	−1.9	−2.4
Technology transfer	High	Moderate	High	Very High
Institutional improvements (world)	High	Moderate	High	Very High
Possible total demand (Gtoe)	Very High 17.2	High 16.0	Moderate 13.4	Low 11.3

reduction in energy intensity are made for all four cases reflecting differing attitudinal assumptions for population groups.

The resulting total demand figures and the breakdown between OECD, CEE/CIS and the DCs is shown in Figure 1.1. It is important to stress that these are not the results of fully fledged scenario studies: they are only the results of some arithmetic exercises based on judgements about possible input factors. Nevertheless, the results carry important messages for energy users, producers and sociopolitical decision makers alike.

The messages from Figure 1.1 are twofold. First, energy demand is likely to continue to grow rapidly. Against median UN statistics for population growth and for three of the four cases, a modest extrapolation of historic global rates of economic growth (3.3% p.a.) but a quickening of pace for the enhanced economic development case (to 3.8% p.a.) and what we believe to be optimistic assumptions about the efficiency with which energy is produced and used to achieve economic growth, energy demand still rises.

It is important to recognize that all four cases assume faster rates of decline in energy intensity – defined as the ratio of gross domestic product to energy use – than have occurred in the past. There is justification for

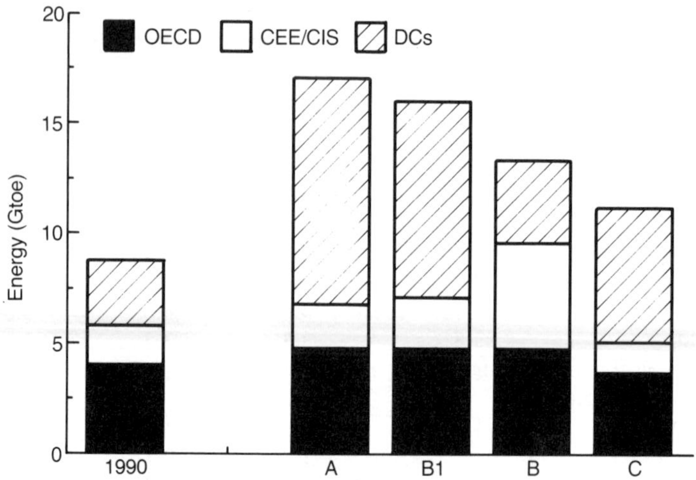

Figure 1.1 Energy demand by group.

this. For instance, economic maturity leads to reduction in energy intensity; there is an increasing tendency to price energy at full cost, including some reflection of the environmental cost, and this will lead to higher efficiency of use; continuing concern in regard to energy use and the environment will maintain pressure on improving efficiency of end use. In any case there is a strong likelihood that the world price of energy - especially of oil and natural gas - will rise in real terms over the next 30 years and beyond. None the less, the range of total demand figures may well be on the low rather than the high side.

In the ecologically driven case rise, in demand is at least contained - but to achieve this, energy intensity across the globe would have to improve by nearly 2.5% per annum which would require fundamental changes in social attitudes and structures.

Second, the developing countries dominate: now, in numbers of people and in growth in those numbers; in the future they will dominate global energy demand and therefore supply. At present, developing countries account for some 33% of global fuel use: the figures suggest in the reference case an increase to over 50% by 2020 and in the enhanced economic development case to 60%. Mention has been made of the implications of population growth for per capita energy consumption - and therefore standard of life - in the developing countries.

The pattern of primary energy use will be heavily influenced by what fuels are available to the population spread across the globe. In developing countries non-commercial fuels are likely to continue to play a vital role in meeting need. The likely pace of growth in non-fossil sources described

Table 1.9 Possible patterns of energy supply (Gtoe)

	1960	*1990*	*in 2020*			
			A	*B1*	*B*	*C*
Coal	1.4	2.3	4.9	3.8	3.0	2.1
Oil	1.0	2.8	4.6	4.5	3.8	2.9
Natural gas	0.4	1.7	3.6	3.6	3.0	2.5
Nuclear	–	0.4	1.0	1.0	0.8	0.7
Large hydro	0.15	0.5	1.0	1.0	0.9	0.7
'Traditional'	0.5	0.9	1.3	1.3	1.3	1.1
'New' renewables	–	0.2	0.8	0.8	0.6	1.3
Total	3.3	8.8	17.2	16.0	13.4	11.3

above suggests that, over the next 30 years or more, fossil fuels will continue to dominate the supply scene. Table 1.9 speculates on possible patterns of supply for the four cases presented. For the reference case commercial coal, oil and gas supply 9.8 Gtoe out of the total 13.4 Gtoe used. Of the extra 3.8 Gtoe needed in the enhanced economic development case, fossil fuels supply 3.3 Gtoe. Of the fossils fuels coal dominates, and will do so increasingly into the longer term as reserves of oil and natural gas decline.

1.3.6 Environmental concerns

An inevitable conclusion from the figures of fossil fuel burn presented is that global emissions of CO_2 are unlikely even to stabilize over the next 30 years or so. Emissions for the four cases are shown in Table 1.10. It will be noted that only in the ecologically driven case are emissions held to near level with the 1990 datum. The situation in regard to CO_2 emissions becomes more interesting beyond 2020.

1.3.7 Beyond 2020

As noted above, the inertia of global energy systems is such that over the next 30 years or so a seemingly inevitable growth in demand is most likely

Table 1.10 Carbon emissions and concentrations for the four cases

Cases	*1990*	*2020*			
		A	*B1*	*B*	*C*
Emissions (GtC)	5.9	11.2	10.1	8.4	6.3
Concentrations (ppmv)	355	434	426	416	404

Table 1.11 Patterns of energy supply to 2100

Case	*1990*	*A*		*B*		*C*	
		2050	*2100*	*2050*	*2100*	*2050*	*2100*
Global energy demand (Gtoe)	8.8	27	42	23	33	15	20
Fossil Fuels (% of primary energy)	77	58	40	57	33	58	15
Nuclear (% of primary energy)	5	14	29	15	28	8	11
New renewables (% of primary energy)	2	15	24	14	26	20	50
Annual CO_2 emissions from fossil fuels (GtC)	5.5	13.7	15.3	11.3	10.8	6.7	2.3

to be met with little radical change in supply technology or dominant fuels. Nevertheless, this period is a vital one in determining whether change is recognized to be necessary and is rigorously pursued in order to meet grave deficiencies in energy distribution and perceived threats to the environment, or whether it is to be business as usual and emergencies met as they appear as best they can be at the time. The demand/supply balance in the period post-2020 takes an interesting turn simply because of the likely decline in availability and escalation in price of oil and gas. Extrapolation of the underlying parameters supporting the three main cases A, B and C through to 2100 yields demands falling within the range 15–27 Gtoe in 2050 and 20–42 Gtoe in 2100, although even these are narrow ranges given how far ahead we are looking (Table 1.11).

The most one might expect fossil fuels to contribute to these global figures would be about 60% in 2050 and in the range 15–40% in 2100. This is based on two premisses. First, by about 2050, with the progressive depletion of reserves (and resources) of oil and gas the prices of these are likely to have eased substantially the competitive position of alternatives. Second, the currently growing concerns at emissions from combustion of all fossil fuels will have intensified the incentives to exploit alternatives such that oil, coal and gas are likely to begin to lose market share in the second half of the 21st century if not before. On that basis, therefore, one might be looking at a nuclear contribution of up to 41 Gtoe by 2050 (compared with the maximum of 1.0 Gtoe postulated for 2020) and up to 12 Gtoe in 2100. That assumes also a parallel rate of development and commercialization of new renewable sources.

1.3.8 Environmental implications

Attention has been drawn to the implications of likely patterns of primary energy supply for CO_2 emissions. Beyond 2020, and on the basis of the figures outlined above, global annual CO_2 emissions are, not surprisingly, likely to continue to rise until the third quarter of the 21st century (Table 1.11). Of greater concern is the prognosis that, based upon work undertaken by the University of East Anglia, using the MAGICC global climate model, atmospheric CO_2 concentrations continue to rise even in the low carbon emission case (Case C) until the last quarter of the next century and well beyond it in the other two cases examined (Table 1.12). A simplistic conclusion is that we have already burnt too much fossil fuel and are too late even now if we find it important to reduce significantly CO_2 concentrations in the atmosphere.

1.4 SOME CRITICAL FACTORS

The simple numeric postulates outlined above underline the point made at the beginning of this chapter that the problems of global energy demand/supply are more than a matter of balancing the numbers. Would that it were as simple as that. At first base the problems are political rather than numeric, socioeconomic rather than technical, even though technology has a vitally important role to play. The major problem is the availability – or lack – of low cost commercial energy to a majority of the world's population, for that deficiency has deep and unpredictable political implications in terms of the aspirations of the developing countries and the relationship between them and their more affluent neighbours in the developed world.

1.4.1 The investment dimension

It is clear that the level of investment required in the developing countries and in the ex-centrally planned economies far outreaches the resources that

Table 1.12 Atmospheric CO_2 concentrations (part per million by volume)

		WEC Commission case			
	IS92A	*A*	*B1*	*B*	*C*
1990	355	355	355	355	355
2020	426	434	426	416	404
2050	523	536	–	494	449
2075	622	615	–	560	475
2100	733	708	–	617	469

can be made available by the aid agencies even if they had the confidence to deploy all that was available to them. The requirements of the developing countries have been estimated at US$1600 billion over the next 10 years, of which some US$1000 billion will be in the power sector. In that sector during the 1980s nearly 90% of the financing came either from tax revenue or from the use of public credit – a mechanism that cannot continue to provide the bulk of a requirement that in the 1990s will be some six times the resources of all the aid agencies combined.

The priority is to create the right conditions for investment whereby the resources of the private sector may be attracted. Although the requirements are immense, the resources overall are even more immense! In 1989 world GDP amounted to US$20 400 billion of which US$4600 billion was for world gross domestic investments. Of this the annual requirement of the energy sector is just 3.4%. Even related to the developing countries and the ex-centrally planned economies alone, the requirements amount to only 8% of domestic investments. Numerically, therefore, and while recognizing the great problems faced today by these emerging economies, the simple figures suggest that the problem is not one of shortage of capital but inability to attract it.

The problem is concerned with the way societies, economies and governments operate as a whole and not just in respect to energy. Sound institutions, political systems, and administrative and economic infrastructures are vital. For instance, a transition to full-cost pricing is seen by many to be essential if adequate finance is to be attracted. This is particularly important in many developing countries. As the demand for energy increases, and as traditional sources of government finance come under pressure, so will it be necessary to seek increasing finance elsewhere. The critical factor is whether companies are seen to be creditworthy; and that will entail their being viable without subsidy and able to generate finance from revenues themselves. This in turn requires sound pricing policies. In practice it is not realistic for developing countries to expect all their financial requirements to be secured on the international markets. The development of indigenous private capital markets must be a goal and many countries are making progress to this end.

Much has been made of the market approach embodying a number of these themes: avoidance of overcentralization of ownership and control; avoidance of *ad hoc* and negative government interference; privatization; introduction of competition and where possible the breaking up of monopolies; establishment of sound pricing and a rational regulatory framework. In theory these must all have important roles to play but in practice the transition is fraught with difficulty as experience over the last few years has amply demonstrated.

Investment in technology, in education and training and in technology transfer between developed countries and developing countries, and between developing countries themselves, must also be seen as vital contributors to meeting the energy deficiency problem. Developing and ex-centrally planned economies will not be able to meet their future energy demand in a sustainable manner without modern higher efficiency technology, which will only be achieved through co-operation with the developed countries. Such co-operation must be based on a fair and balanced transaction between a motivated technology buyer/receiver and a motivated, informed technology seller/supplier; and for this to happen there must be an incentive policy and the right institutional environment. Given these, co-operative development via free trade, foreign investment, joint ventures and equity participation consistent with honouring reasonable intellectual property rights must follow.

1.4.2 Never has so much been needed by so many from so few ...

A perplexing conclusion from the geographic pattern of demand and supply into the 21st century is that more and more people, in absolute and in percentage terms, will become dependent for their energy supplies on fewer and fewer people. Table 1.13 illustrates the point. Table 1.14 shows that although coal resources are comparatively well balanced across the globe, and traded coal represents, now, only about 10% of consumption, the imbalance in oil and gas is much more pronounced. In 1990 the Middle East accounted for 60% of oil movements; the CIS accounted for 47% of gas movements. World trade in oil accounted for just under 50% of consumption; just over 30% of fossil-fuel-based energy was traded across borders. Roughly a quarter of the world's population is entirely dependent upon imported energy. In fact the world can be divided up into energy rich exporters, comprising in 1990 just over 40 % of the population; energy rich importers, comprising just over 30% of the population; and energy poor importers, comprising just over 20% of the population. By 2020 that pattern will have changed dramatically. The 30% fossil energy traded across

Table 1.13 Estimated population of net energy importing countries (millions)

	1990	*2020*
Industrialized countries	700	1000
Economies in transition	100	0[a]
Higher income developing countries	450	1100
Lower income developing countries	1650	5100

[a]Economies in transition are classed as industrialized countries by 2020. However, the bulk of the population in this category is defined as energy exporters. (Source: WEC)

Table 1.14 Distribution of fossil fuel reserves (% of world proved total)

	Coal	*Oil*	*Natural gas*
Asia	20	4	6
Australasia	9	–	–
Middle East	–	63	34
CIS	23	18	40
CEE	7	–	–
OECD Europe	10	1	4
N. America	24	4	6
Latin America	2	10	6
Total	95	100	96

Source: WEC Survey of Energy Resources.

borders in 1990 will be nearer 50% in 2020. The 40% energy rich exporters will have shrunk to just over 10%; the 30% energy rich importers will have risen to over 40% and the 20% energy poor importers to over 45%.

It is easy enough to conclude that to be concerned at postulated numerical patterns of this nature is to be alarmist in the extreme. After all, we do get on reasonably well with everybody else, don't we? While, certainly, problems arising from energy supplies have in recent history been confined to local areas – including the Middle East in that definition – it would be as well to consider what measures ought to be pursued in order at least to alleviate potential difficulties. These must include diversifying imported fuel sources and seeking to increase indigenous supplies; improving efficiency of end use and improving institutional and co-operative arrangements between energy suppliers.

It is important to recognize also that while the non-CO_2-emitting technologies such as nuclear power and the renewables are the obvious and important counters to the threat of global warming, it is not axiomatic that they can be exploited in the right places to achieve the desired result. For instance, over the next 30 years China is likely to increase its fossil fuel burn from around 700 Mtoe in 1990 to around 1550 Mtoe in 2020: that represents some 30% of the global increase in fossil fuel burn 1990–2020 for the reference case. The World Energy Council Commission (1993) postulates an increase in the nuclear capability in China from zero to 40 Mtoe in the same period – about 5% of the increase in fossil fuel burn. While the developing countries as a whole are assumed to make an increasing contribution to global nuclear output only about 30% of that increase can reasonably be apportioned to them in the period 1990–2020. But these countries account for roughly 75% of the increase in global energy requirement over the same period and for 85% of the increase in fossil fuel burn.

1.5 SOME CONCLUSIONS

The seemingly inevitable growth in the demand for energy can be met by continued exploitation of fossil fuel reserves for many decades into the 21st century. However, an apparent balance of the global demand/supply equation through exploitation of all primary energy supply technologies masks the real threats imposed by global warming, continuing imbalance in energy availability across world populations and, particularly in the case of nuclear power and renewable energy forms, how these technologies are to be exploited to the benefit of the poorer developing countries. The primary characteristics of the global energy demand/supply situation are summarized below. It is a very broad picture and contains some elements barely touched upon, if touched upon at all, in this chapter. Nevertheless, it is hoped it provides a useful background against which to consider the role of energy technology in the third millennium.

1. A majority of the world's population does not have enough energy or the services it provides. This is the primary problem to be tackled. Energy strategy must have as its primary objective the provision of adequate, affordable and reliable energy services, available to all people.
2. However, world energy problems and perspectives are not uniform across the industrialized, developed countries, the economies in transition and the developing countries. Energy policies must be designed to respond to the specific needs of these countries and groups of countries and enhance trade and co-operation between them in order to capitalize on complementarities. There are no universally applicable solutions.
3. In developed countries, improvements in energy efficiency are likely to continue, enabling increase in energy demand to be contained such that growth in energy use will be less than growth in the economy, and possibly negative. The problems here are best dealt with by market liberalization wherever there are distortions or constraints. Fossil fuels will remain predominant. In these countries, energy policies should continue to drive to enhance the efficiency of energy use and accelerate the diversification of the supply base.
4. In the economies in transition, the main issues are the modernization of the energy supply infrastructure, the promotion of the rational use of energy and the transition to market oriented policies and enterprises. International co-operation strategies should enhance the exchange of experience on policy, legal, institutional, economic and technical issues of transition.
5. In the developing world, population growth will mean rapid increase in total energy but low increase in per capita consumption. To the extent that demand may be, as at present, largely met by conventional fuels and technologies, demand growth will be associated with

increasing environmental hazard. But energy growth must occur if economic improvement is to be achieved. There are two imperatives: first, to identify how energy demand is to be met and, second, to contain environmental pollution through clean technology and increased energy efficiency. It is imperative that in these countries energy strategies drive first and foremost to meet demand, but that technology and supply patterns are improved, particularly through the introduction of clean technologies and the promotion of the rational use of energy.

6. There is no global shortage of primary energy as such, nor is there likely to be, although reserves are unevenly distributed. The real constraint in providing energy to the majority in need is the inability to attract capital. Massive investments are required. International and domestic capital markets are large enough to supply efficiency-driven energy systems but local institutions and investment environments do not encourage economic investment. Restructuring of institutional frameworks and establishment of policies that will enable developing countries to attract finance and procure technology is vital; full cost pricing systems should be adopted over time as soon as political, social and economic circumstances permit; open market competition, and encouragement of equity capital investment through international energy enterprises, are needed. Governments should set proper framework conditions, leaving energy enterprises to energy operatives within a framework of accountability.
7. International trade in energy and technology will be required at increased levels. Protectionist barriers, supply monopolies and subsidized energy lead to market distortions which hinder economic development in poorer countries and global improvement in energy use. Those who lack natural resources or technology must be able to trade for them. Interdependence in all aspects of energy supply and use – fuel, technology, finance and know-how – must be made to work through regional and international global co-operation, the development of market systems and of private investment across national boundaries, and by open access to markets for trade.
8. All energy production, transportation, conversion and use have an environmental impact. The wealthy nations have the resources and technology to resolve the majority of their local environmental problems and to face up to broader, global issues. Many of the poorer nations have yet to deal not only with the local effects of energy use but also with other externalities such as clean water, sanitation, etc. For this group of nations these problems, and energy provision, take precedence over global environment concerns. Ways to avoid the costly environmental repercussions of energy use in these countries have to be found; much can be achieved through existing technology.

Governments, consumers and the energy industries should ensure that energy supply and use are compatible with protection and preservation of the environment by establishing economic ground-rules against which efficiency improvement may be assessed, by setting up effective independent regulatory processes and by making the polluter pay by including the cost of environmental protection in energy prices.

9. Conservation and a continuing improvement in the overall efficiency of energy provision and use are essential both to increase economic wealth and to reduce the burden on the environment, although it is recognized that their application in many regions of the world can only be achieved to the full by a competitive market place economy. Governments and the energy industries must continue to emphasize the potential economic value of improving efficiency through advising, promoting and informing. Competition in supply and prices that fully reflect costs, including environmental costs, are important incentives to increasing efficiency.
10. Over the next few decades fossil-based energy sources are likely to continue to provide the bulk of world needs; but all primary sources that present the prospect of economic exploitation need to be developed. It is probable that within the second quarter of the 21st century, and perhaps before, the extent of depletion of the less expensive oil and gas resources will have begun to reflect on price. Alternative technologies, new and renewable energy forms and nuclear energy will be required more urgently. It is now – and not in 20 or 30 years' time – that governments must provide adequate R&D support to new technologies and assist them to achieve technological and commercial viability if, because of the timescales involved, the market does not do so.
11. Given the priorities of economic growth and the dominance of fossil fuels, it is likely that carbon dioxide emissions will continue to rise significantly world-wide well into the next century. Environmental and social policy should be realistically based on this expectation, but governments should actively promote development and commercial exploitation of economic contributions from non-fossil energy sources, especially from the renewables and nuclear energy; and co-operate in developing cleaner applications of fossil fuels.

ACKNOWLEDGEMENTS

The author gladly pays tribute to the many colleagues with whom the work underlying this chapter was undertaken on behalf of the World Energy Council, while the author was Executive Director of the Council's Commission 'Energy for Tomorrow's World'. Nevertheless the views expressed in this chapter, while resting heavily on the work of the Commission, are those of the author and not necessarily those of the World Energy Council. The report of the Commission is now published (World Energy Council, 1993) and readers are invited to refer to it.

REFERENCE

Energy for Tomorrow's World – the Realities, the Real Options and the Agenda for Achievement (1993). The World Energy Council. Kogan Page, St Martin's Press.

2

Costing the environmental damage from energy production

D. Pearce

2.1 INTRODUCTION

It is now widely understood that the production and use of energy are inefficient because of the failure of energy prices to reflect the full costs of production and use. Full costs equal private costs (capital, operation and maintenance, R&D) plus external costs. External costs are costs imposed on local, national or global communities and which go uncompensated in the market system. External costs comprise environmental and non-environmental costs such as global warming impacts from CO_2 and CH_4, acid rain, energy security risks, hidden subsidies, etc. This chapter looks at the likely size of environmental externalities in the United Kingdom and the way in which they would affect energy prices.

2.2 SOCIAL COSTS AND ENERGY PRODUCTION

All energy production yields benefits in the form of deliverable heat, light and power. As a very rough first approximation, we can measure the benefits of energy supply by the price that is charged for it in the market place. This measure assumes that energy is sold in free markets, i.e. that markets are not distorted by subsidies. Where such subsidies exist, and we know they are extensive in both the OECD and the developing world, careful adjustment has to be made to estimate the true benefits.

All energy production also imposes costs on society. Those costs take two forms. The first is the cost of producing and delivering the energy. This is what most people usually understand by the costs of supply or costs of production. The second form of cost arises from the fact that energy production produces unwelcome side effects. This may be the visual intrusion or noise associated with wind power farms. It may be the acidic pollutants and particulate emissions associated with coal-fired electricity, the risks of health damage from routine radiation, the risk of a nuclear accident, the damage caused by a tanker oil spill, the impacts on biological diversity caused by tidal barrages, the risk of explosion from gas exploitation, and so on. Somewhat confusingly, this second category of costs goes under various names: they are known as externalities, external costs, neighbourhood effects and third party effects. More recently, they have become known in the energy literature as 'adders' – because they are additional costs to the costs of production. The important thing about these additional costs is that, in the absence of regulation, they will be ignored by the producer of energy. The builder of the tidal barrage will not be concerned with the displacement of bird feeding grounds for example. The coal-fired power station owner will not be concerned with the acid rain emissions he causes, emissions that become depositions both in the country of origin and, often, abroad as well (the phenomenon of 'transboundary' pollution). Similarly, the owner of a fossil fuel boiler will not be concerned with the carbon dioxide emissions that contribute to the possibility of global warming.

In an unregulated world, then, energy prices will be too low. For they will reflect the costs of production, including the desired profit of the energy producer. But they will not reflect the externalities. Yet the externalities are as 'real' a cost as the costs of production. This is because cost in economics relates to any loss of individual well-being (economists tend to use the terms welfare or utility). Therefore an unregulated, free market system will fail to achieve the most desirable allocation of energy resources. Not only will too much energy be consumed, the mix of energy sources will also, very probably, be wrong.

In reality, most of these externalities are regulated in one way or the other. There are national regulations, often based on the idea of adopting the 'best available technology' (BAT) to control emissions. There are regional regulations, such as the Large Combustion Plants Directive of the European Union. There are wider international regulations, such as the Sulphur and Nitrogen Protocols under the Convention on the Long Range Transport of Air Pollution. And now there is even a global restriction on carbon dioxide under the Framework Convention on Climate Change, negotiated at Rio de Janeiro in 1992, and implemented in March 1994. The existence of these regulations means that we do not, in practice, allow free rein to market forces. We regulate the externalities. The issue is, do we regulate them as much as we should? If not, then one way of providing additional regulation to reflect externalities is to impose some form of pollution tax.

There are also taxes on energy sources, taxes that invariably have nothing to do with the externalities, but which exist to raise revenue for the government. These taxes raise the price of energy, and reduce its consumption. However, these taxes are rarely related to the pollution impact of energy production and hence do not produce the right mix of energy sources. Moreover, although the arguments are quite complex, environmental taxes should be on top of, not instead of, general revenue raising taxes (Newbery, 1992).

2.3 THE THEORY OF EXTERNALITY TAXATION

If externalities are ignored in the production and use of energy, then the allocation of resources in the economy will be inefficient. Since the price of any given energy resource will tend to reflect its cost of production, energy with low costs of production will tend to be used first. Then the next most expensive source will be employed, and so on. Indeed, this is exactly how electricity systems operate. But suppose the cheapest source of energy happens also to be the most polluting. Then the market mechanism will bring that source into production despite the external costs associated with it. The correct mix of energy production should be determined by the production costs plus the external costs. This grand total is usually also known as the social cost of production. The basic equation is therefore:

$$\begin{aligned}\text{Price} &= \text{Marginal Production Cost} + \text{Marginal External Cost}\\ &= \text{Marginal Social Cost}\end{aligned}$$

The term 'marginal' simply means 'extra'. Figures 2.1 and 2.2 illustrate the basic concepts. The aim of policy should be to raise the marginal production cost curve to the level of the marginal social cost curve. This can be done with a tax based on the pollution damage caused, or by using a (tradeable) emission permit system, or, of course, by direct regulation such as the setting of an environmental standard. For reasons not pursued here, economists tend to argue that the use of tradeable permits or environmental tax schemes (market-based instruments) is more efficient than the use of direct regulation. Basically, market-based instruments have the capacity to yield the same amount of environmental quality, but at lower cost to industry. They minimize compliance costs (Pearce and Turner, 1990).

If the tax is to be related to the amount of damage done by each energy source, then Figures 2.1 and 2.2 raise an immediate problem. The tax is in money terms, i.e. money per kWh or per BTU, etc. The production costs are also in money terms. To relate the tax to damage therefore requires that we also estimate the pollution damage in money terms. (This is not a strict requirement. However, it will not be possible to achieve the most desired position, the 'social optimum' in Figures 2.1 and 2.2, unless we know the money value of damage.) This then takes us into the problem of placing money values on the pollution damage, or as it is often, but misleadingly, termed, 'valuing the environment'.

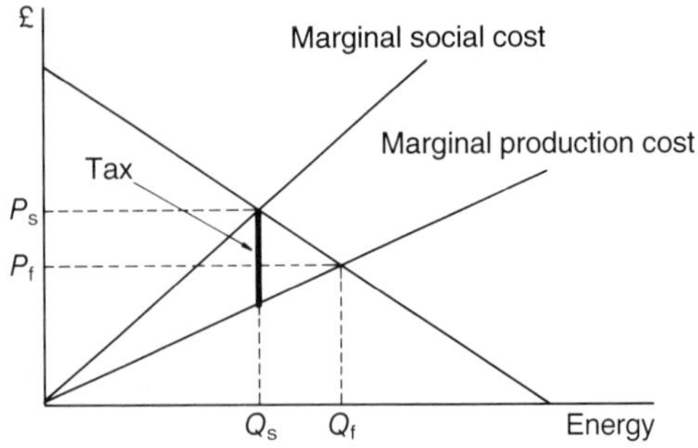

Figure 2.1 Illustration of marginal costs.

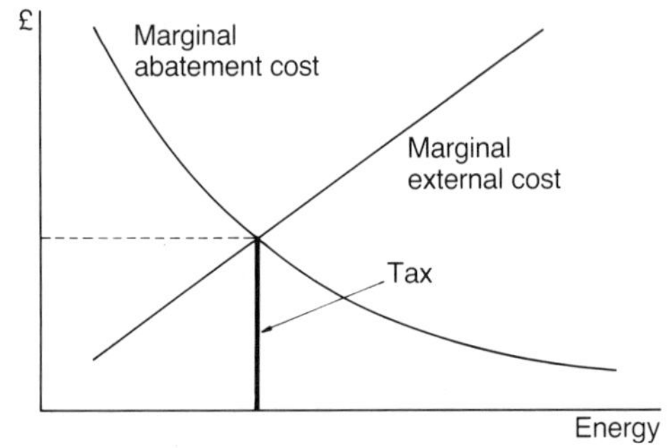

Figure 2.2 Marginal costs of energy production.

The rationale for finding monetary values of externalities is as follows. The costs of production of energy are expressed in money terms. The cash value of the resources used to produce energy reflects the scarcity of those resources. Generally, the scarcer the resource relative to the demand for it, the higher the price. So, production costs tell us something about society's perception of scarcity. The prices tell us something about the value of those resources if they were used in some alternative way, what economists call the 'opportunity cost'. In turn, that opportunity cost reflects society's valuation of the benefits of that alternative use, a valuation that reveals itself in the willingness to pay for the benefits yielded by those resources. So, the first observation is that production costs themselves reflect the willingness of society to pay for the benefits forgone by using scarce resources in energy production.

The link between willingness to pay and the benefit of energy production and use is more direct. The demand curve in Figure 2.1 shows how demand varies with price. The demand curve is in fact a 'willingness to pay curve' (more strictly, it is the marginal willingness to pay curve, i.e. the willingness to pay for one extra unit of energy). In choosing whether to pay a given price, consumers weigh up the benefits of the energy source. The demand curve traces out the benefits that consumers expect to get for each quantity of energy.

The monetary value of the externality, then, is also a reflection of society's willingness to pay, this time for avoiding pollution damage, or the amount of money that is required by way of compensation to tolerate the damage.

In this way, all the components of Figure 2.1 - demand, costs of production and environmental impacts - are linked through the concept of willingness to pay. When economists 'value the environment', then, they are not engaged in some strangely immoral behaviour. They are measuring people's preferences for or against environmental change. In doing this they are seeking to allocate environmental resources in the same way that many other goods and services are allocated - according to market values. It is perfectly reasonable to argue that the environment should not be allocated in this way. Perhaps it is, in some sense, sacrosanct. If so, we might develop other rules for its allocation. But what those rules are is not obvious, and any rule faces as many problems as the willingness to pay criterion (Pearce, 1993).

This discussion enables us to proceed to the issue of placing actual money values on environmental damage from energy production, i.e. to estimate 'externality adders'. As it happens, environmental impacts are not the only kind of 'adder' that is relevant. There are many non-environmental impacts that give rise to distortions in energy markets. For example, some sources of energy may need to be imported from areas of the world where there is risk of political disturbance. If resources are spent on, for example, military installations in the area in an effort to protect those resources, then those expenditures can rightly be thought of as extra costs of that energy supply. Yet the military expenditures do not get reflected in the price charged for the resource, e.g. oil. This is a further distortion: the oil price is effectively being subsidized by military expenditures.

2.4 HOW ARE ADDERS ESTIMATED?

Estimating adders involves some monetary estimate of the externality. In the case of energy security, for example, it might be the cost of maintaining armed forces in the Middle East or in readiness for use in that area. In the case of environmental externalities, monetary estimates involve the still controversial but now widely accepted procedure of estimating the willingness to pay for avoiding or cleaning up damage. This might be done by asking people (contingent valuation), by observing expenditures on travel to recreational sites (travel cost method), by looking at the way property prices respond to differences in environmental quality (hedonic property price method), and by using market or near-market values to value physical impacts such as crop losses, forest damage, buildings damage, etc. (the dose–response or production function approach).

Money estimates for adders to electricity systems have been estimated in a number of studies (e.g. Pearce, Bann and Georgiou, 1992; ETSU 1993; Resources for the Future, 1992). We describe briefly below a procedure for estimating adders for the UK electricity system. These estimates develop those in Pearce, Bann and Georgiou (1992) and Pearce (1993). More details for some of the pollutants can be found in CSERGE, EFTEC and Warren Spring Laboratory (1994).

Pearce, Bann and Georgiou (1992) is typical of what might be called 'top-down' approaches in which some estimate of national damage is divided by total pollutant depositions to obtain a damage per unit of pollutant. These top-down approaches may be contrasted with 'bottom-up' approaches in which damage from a single emitting source - actual or hypothetical - is traced or simulated (ETSU, 1993; Resources for the Future, 1992). Both procedures have shortcomings. The top-down approaches clearly obscure impacts in one overall average. Bottom-up approaches have little relevance for national damage estimates unless the single source can be regarded as being very typical of all such sources.

The procedure for estimating adders is first illustrated for conventional air pollutants. Results for carbon dioxide and nuclear power are reported only briefly.

2.5 THE ADDERS FOR THE UNITED KINGDOM: FOSSIL FUELLED ELECTRICITY

2.5.1 Health

(a) *Mortality*

Pearce, Bann and Georgiou (1992) estimate health damage by the electricity fuel cycle, but do not allocate the costs to pollutants. The following procedure yields an approximation of damage in pence per kilogram (p/kg) of pollutant.

Damage costs for **mortality** are given as:

$$D_c = 0.32\,\text{p/kWh for coal-fired electricity}$$
$$D_o = 0.29\,\text{p/kWh for oil-fired electricity}$$
$$D_g = 0.02\,\text{p/kWh for gas fired electricity}$$

We compute total mortality costs in the United Kingdom due to fossil-fuel electricity (f) as:

$$D_f = \Sigma D_i \cdot Q_i$$

where Q_i is the output of the ith source of electricity and D_i is the damage from the ith source. For 1991

$$Q_c = 195\ 000\ \text{GWh},\ Q_o = 23\ 000\ \text{GWh},\ Q_g = 0$$

so that

$$D_f = \Sigma D_i \cdot Q_i = £624\ \text{million} + £67\ \text{million} = £691\ \text{million}$$

Using shares of 13%, 50% and 37% to allocate damage to pollutants (Norway CBS, 1991), the resulting damages are:

SO_2	£90 million
NO_x	£345 million
TSP	£256 million

Two adjustments are necessary. First, we divide by (1990) total emissions from power stations to obtain:

$$SO_2 = £90\ \text{million}/2.72\ \text{million/tonnes} = £33/\text{tonne} = 3.3\ \text{p/kg}$$
$$NO_x = £345\ \text{million}/0.78\ \text{million/tonnes} = £442/\text{tonne} = 44.2\ \text{p/kg}$$
$$\text{TSP} = £256\ \text{million}/0.027\ \text{million tonnes} = £9481/\text{tonne} = 948.1\ \text{p/kg}$$

The second adjustment makes allowance for the fact that emissions and depositions differ due to the long range transportation of SO_x and NO_x. The figures above are for damage per kilogram deposited. A unit of emission of sulphur, for example, will result in a fraction of that emission being deposited in the United Kingdom. The remainder will be deposited in the sea or on other countries. Assuming damage to other countries is ignored, we multiply by a further factor of 0.36 for SO_2 and 0.13 for NO_x to allow for the ratio of depositions to emissions (CSERGE, EFIEC and Warren Spring, 1993) to obtain:

$$SO_2 = £11.9/\text{t} = 1.19\ \text{p/kg}$$
$$NO_x = £57.5/\text{t} = 5.75\ \text{p/kg}$$
$$\text{TSP} = £9481/\text{t} = 948\ \text{p/kg}$$

The 'value of a statistical life' used in Pearce, Bann and Georgiou (1992) is £2 million. While this is warranted in light of a review of the literature, it is above that used in current UK government practice. A figure of £0.7 million would be consistent with Department of Transport practice, for example.

Assuming unit deposition damage gives rise to the same levels of health damage in Europe generally, the unit values of 3.3 p/kg for SO_2 and 44.2 p/kg for NO_x need to be multiplied by 0.56 for SO_2 and 0.51 for NO_x to obtain European health damage arising from UK emissions. Note the rationale for including these wider geographic damages is that the United Kingdom is party to a number of European environmental agreements on transboundary air pollution. Europe is defined here as the UN Economic

Commission for Europe region, i.e. it includes Scandinavia and the East European countries, as well as the western countries of part of the old Soviet Union. This gives the following results for UK adders:

$$SO_2 = 1.85\ \text{p/kg}$$
$$NO_x = 22.5\ \text{p/kg}$$
$$\text{TSP} = 948\ \text{p/kg}$$

(b) *Morbidity*

In Pearce, Bann and Georgiou (1992) damage costs for morbidity are given as 0.12 p/kWh and are taken from PACE (1990), a study relating to the United States. These costs are directly attributable to particulates (TSP). This figure translates to £7.5/kg of TSP. More recent work by Brajer, Hall and Rowe (1991) and Hall *et al.* (1992) suggests that morbidity damage from TSP is likely to be less than mortality damage. Evaluating the impacts of the control of TSP and ozone in the South Coast Air Basin centred on Los Angeles, they conclude that restricted activity days (RADs) and illness symptoms have an economic value of 50–70% of mortality gains. Since their work is the most thorough analysis to date we take the 50% value here, but the use of such figures raises the general issue of 'benefits transfer', i.e. the extent to which damage and benefit estimates in one context can be transferred to another context.

It is worth noting that the estimates here for health damage are very considerably below those quoted in ECOTEC (1992) for health damages. ECOTEC's estimates suggest something of the order of 185 p/kg (£1.85/kg) damages.

2.5.2 Crop damage

Pearce, Bann and Georgiou (1992) use a National Acid Precipitation Assessment Program (NAPAP) estimate for the United States equivalent to 20 p/kg of NO_x for crop damages on the assumption that ozone damage can be ascribed to NO_x. The only adjustment required here is that which allows for the transboundary nature of NO_x. Multiplying by 0.13 produces a damage figure for the United Kingdom of 2.6 p/kg and by 0.51 gives 10.2 p/kg for Europe.

ECOTEC (1992) suggest a UK figure of some 10 p/kg.

2.5.3 Forests

The damage figure in Pearce, Bann and Georgiou (1992) of 60 p/kg of SO_2 deposited comes from a study of forest damage in Europe by Nilsson (1992). This study is controversial in that it ascribes damage to SO_2 alone, whereas the role of acidic pollutants in forest damage is disputed – see especially NAPAP (1990). It also takes a fairly arbitrary multiplication factor of 2.7 of commercial timber values to obtain a total forest value inclusive of recreational values. Multiplying by 0.36 and 0.56 for the emissions/deposition ratio relevant to the UK and European damage gives 22 and 33.6 p/kg respectively. We reduce this further to 10 p/kg for the United Kingdom and 20 p/kg for Europe to lower the exaggerating effect of the recreational use 'multiplier' (Pearce, 1990). This figure remains controversial.

ECOTEC (1992) suggest a figure of around 73 p/kg.

2.5.4 Buildings

The original damage estimate in Pearce, Bann and Georgiou (1992) was based on provisional estimates by ECOTEC of buildings damage in the United Kingdom. Since the publication of Pearce *et al.* ECOTEC have revised their estimate of damage downwards in a significant way, from some £2.30 per kg SO_2 deposited to some £0.37 per kg (ECOTEC, 1992). Taking this at face value, and multiplying by 0.36 and 0.56 to go from deposition to emissions gives a figure of 13.3 p/kg for the United Kingdom and 20.7 p/kg for Europe.

2.5.5 Water

Pearce, Bann and Georgiou (1992) were unable to produce estimates for water damage.

The results of section 2.5 (rounded) are summarized in Tables 2.1 and 2.2.

Table 2.1 Average unit damage costs, UK (p/kg)

	SO_2	NO_x	*TSP*
Health:			
Mortality	1.2	5.7	948.0
Morbidity	–	–	474.1
Crops	–	2.6	–
Forests	22 adjusted to 10.0		–
Buildings	13.3	–	–
Water	N/A	N/A	N/A
Total	24.5	8.3	1422.1

Table 2.2 Average unit damage costs, ECE (p/kg)

	SO_2	NO_x	*TSP*
Health:			
Mortality	1.8	22.5	948.0
Morbidity	-	-	474.1
Crops	-	10.2	-
Forests	33.6 adjusted to 20.0		-
Buildings	20.7	-	-
Water	N/A	N/A	N/A
Total	42.5	32.7	1422.1

2.6 CONVERSION TO ADDERS

To convert these figures to 'adders' we need to know the relationship between emissions and electricity produced. Table 2.3 provides information on emission coefficients. They relate to the whole fuel cycle and not just electricity generation. They are based on ETSU data (Eyre, 1990). Table 2.4 shows the adders in p/kWh for selected main sources of electricity generation - coal, gas and nuclear. The effect is clearly to militate against conventional coal-fired plant, i.e. plant without gaseous emission control. The adder for a new coal-fired plant, i.e. with pulverized fuel and fuel gas desulphurization, is seen to be very much smaller.

In this way we can build up a picture of the adders that are relevant to

Table 2.3 Emission factors for UK fuel cycles (grams emissions/kWh)

Cycle	SO_2	NO_x	*Part*	*GHGs as Cequ*
Old coal	14	5.3	0.16	352
New coal	1.2	2.7	0.16	296
Old oil	16.4	2.5	0.16	300
Gas	0.5	0.9	?	139
Nuclear	0	0.2	0	14
Landfill	0.5	0.9	?	139
Waste	4.1	2.7	6.4	?
CHP	0.6	1.3	0.08	150
Renewables	0	0	0	0

Source: Adapted from Eyre (1990). Estimates for the nuclear fuel cycle are adapted from Table 8 of Eyre (1990) on greenhouse gas emissions; CO_2 emissions from the nuclear fuel cycle are disputed. This controversy relates mainly to the pollution impacts from the *construction* of the generating station. No account is taken here of the relative emissions from plant construction other than through human health effects.

Table 2.4 Illustrative adders for 'conventional' air pollution in the UK and ECE region[a] (p/kWh)

	SO_2	NO_x	*TSP*	*Total*
Old coal				
UK	0.34	0.04	0.23	0.61
ECE	0.59	0.16	0.23	0.98
New coal				
UK	0.03	0.02	0.23	0.28
ECE	0.05	0.08	0.23	0.36
Gas				
UK	0.01	0.01	neg	0.02
ECE	0.02	0.03	neg	0.05
Nuclear	zero	neg.	zero	neg.

[a] ECE includes the UK.

each electricity system. Pearce, Bann and Georgiou (1992) estimate adders for carbon dioxide emissions, i.e. global warming damage, and radiation hazards.

2.6.1 Global warming

Since publication of Pearce, Bann and Georgiou (1992) further and more credible estimates of global warming damage have been provided by Fankhauser, reported in Fankhauser and Pearce (1993). These have the effect of raising the CO_2 damage estimates from £5.8 to £17.3 per tonne of carbon to a single best estimate of £14 per tonne of carbon. (Pearce, Bann and Georgiou (1992) used an average figure of £11.55 per tonne of carbon in their summary tables. The adjustment reported here, therefore, raises the adder by a factor of 14/11.5 = 1.2.) These estimates are based on assessments of likely economic damage from sea level rise and other impacts of warming on crops, forest, etc. Recent scientific work also suggests that sulphur aerosols may actually have a cooling effect on global climate, offsetting to some extent the effects of the warming 'greenhouse' gases. If so, then, strictly, sulphur emissions have a global warming benefit. No account of this is taken here but it clearly should enter into future extensions of the adders work (Maddison, 1994).

2.6.2 Nuclear radiation

Radiation hazards are problematic since human health risks for routine radiation result in almost negligible adders for nuclear power. But the public's perception of radiation hazards from a major accident, especially since Chernobyl, means that some account must be taken of the anxiety and concern shown. The objective risk versus perception issue turns out to be important. Consider the objective risk of a nuclear power plant accident in the United Kingdom. Taking design risk factors for new PWRs such as that planned for Hinkley Point, and multiplying by the £2 million life value gives a 'nuclear accident adder' of 0.0000125 p/kWh, which is negligible. However, these risks are for accidents in which tens or hundreds of deaths occur because of the one accident. There is evidence to suggest that people do not value risks of group accidents in the same way as they value individual deaths. Indeed, casual empiricism suggests this is so since, of the many road deaths each year, those that are reported by the media tend to involve only multiple deaths. This suggests that there should be some multiplication factor for 'group accidents' reflecting 'disaster aversion'. But it is open to debate what the factor is and there appears to be no consensus in the literature.

The number of exceeded fatal cancers per reactor year for a degraded core accident is taken from the National Radiological Protection Board. These give the second row in Table 2.5 (f for frequency, n for number).

The last three rows of Table 2.5 show how various disaster aversion functions might be used to reflect the fact that society tends to weight group losses more heavily than single deaths. The 'square rule' (row 3) is suggested by several authors but it is not based on empirical studies (none the less, one public utility is known to use it in its safety planning).

Analysis of the available evidence suggests that the risk of low probability, high consequence PWR accidents may be valued comparably with as much as about eight individual deaths per reactor year. At £2 million per statistical life, this would produce an adder of £16 million across 6000 million kWh per GWyr, i.e 0.27 p/kWh.

Row 4 in Table 2.5 uses a disaster aversion function suggested by Rocard and Smets (R–S). Self-evidently, the R–S rule gives higher values for 'damage' up to a value of $n \times 300$. After that, the R–S rule gives lower damage values compared to the square function rule. If one reworks Table 2.5 with $f \cdot 300n$ instead of the square rule, then the lives lost are very few,

Table 2.5 Disaster risk factors

$f =$	10^{-8}	10^{-9}	10^{-10}	10^{-11}
$n =$	11 000–35 000	46 000–150 000	110 000–350 000	180 000–580 000
fn^2	1–12	7–22	4–12	1–3
$300fn$	0.03–0.11	0.01–0.05	0.003–0.01	0.0005–0.002
$f.n^{3/2}$	0.01–0.07	0.01–0.06	0.004–0.02	neg.–0.004

the highest value would be £0.21 million and the lowest £1000 or so. At £0.2 million the adder would be negligible at 0.0033p/kWh, and at £1 million would be 0.016 p/kWh. On the other hand, R–S is linear in *n* and this seems to offend the general intuition in the risk aversion literature. Looking at non-linear functions, we illustrate a further possible function of *n* to the power 3/2. This produces similar small adders to the R–S rule.

Overall, then, a disaster aversion adder remains very uncertain. Use of a square function could make the adder as high as 0.67 p/kWh, but the square function has no apparent empirical basis. Use of a Rocard–Smets function, which is linear in the number of people in the group accident, produces adders of 0.02p/kWh and perhaps as high as 0.05p/kWh. The issue is unresolved in the absence of fully fledged risk perception studies for the United Kingdom.

As far as nuclear accident costs are concerned, the health damage costs noted here do not take into account any property and output losses, nor any other economic costs of land sterilization.

It is also important to note that major accidents occur in other fuel cycles, notably with coal (mining accidents), oil and gas (offshore disasters, gas terminal explosions), and hydropower (dam bursts). These were not estimated in Pearce, Bann and Georgiou (1992).

2.7 SOME ELECTRICITY ENVIRONMENTAL ADDERS FOR THE UNITED KINGDOM

Table 2.6 shows the results of combining the global warming and nuclear adders with those for conventional air pollution. The ranking is very much as one would expect: existing coal attracts the highest adder, new coal next, then nuclear and gas in equal positions if allowance is made for disaster

Table 2.6 Provisional adders for the UK (p/kWh)

	Old coal	*New coal*	*Gas*	*Nuclear*
Conventional				
Air pollution				
UK damage only	0.62	0.28	0.02	neg.
All ECE damage	0.98	0.36	0.05	neg.
Global warming	0.48	0.41	0.19	0.01
Routine radiation	0.02	–	–	–
Disasters	?	?	?	0.02–0.27
Sum				
UK only	1.11+	0.69+	0.21	0.03–0.27
All ECE	1.48+	0.77+	0.24+	0.03–0.27[a]

[a] Assumes no contamination of countries other than the UK in the event of an accident.

aversion, and in the order gas then nuclear if aversion is ignored. But, of course, there are many caveats that can only be minimized with further research. We have omitted disaster costs from coal and gas; all the environmental damages are uncertain; non-environmental externalities are not estimated at all; the top-down approach probably exaggerates adders anyway; future private costs of generation are themselves uncertain, and so on. But the estimates do perhaps narrow the policy debate somewhat.

REFERENCES

Brajer, V., Hall, J. and Rowe, R. (1991) The value of cleaner air: an integrated approach. *Contempororary Policy Issuses*, **IX**, April, 81–91.

CSERGE, EFFEC and Warren Spring Laboratory (1994) *Externalities from Landfill and Incineration*. HMSO, London.

ECOTEC (1990) *A Cost Benefit Analysis of Reduced Acid Deposition: a Revised Approach Evaluating Buildings Materials*, ECOTEC, Birmingham.

ETSU (1993), Report to the European Commission on Social Costs of Energy. CEC, Brussels 1992.

Fankhauser, S. and Pearce, D.W. (1993) The social costs of greenhouse emissions. Paper to OECD/IEA Conference on the Economics of Climate Change, Paris, June 1993.

Hall, J., Winder, A., Kleinman, M., *et al.* (1992) Valuing the health benefits of clean air, *Science*, **255**, 14 February, 812–17.

Maddison, D. (1994) The shadow price of sulphur aerosols as global temperature coolants. Centre for Social and Economic Research on the Global Environment (CSERGE), University College London.

NAPAP (National Acid Precipitation Assessment Program) (1990) Integrated Assessment Report. NAPAP, Washington DC.

Newbery, D. (1992) Should environmental taxes be added to existing excise taxes? Department of Applied Economics, University of Cambridge.

Nilsson, S. (1992) Economic impacts of forest decline caused by air pollutants in Europe. *The Economic Impact of Air Pollution on Timber Markets*. US Department of Agriculture, Asheville, North Carolina.

Norway Central Bureau of Statistics (1991) *Rapporter 91/1A: Natural Resources and the Environment 1990*. Oslo.

PACE (1990) *Environmental Costs of Electricity*. Oceana Publications, New York.

Pearce, D.W. (1990) *Forestry Expansion: A Study of Technical, Economic and Ecological Factors*. Forestry Commission, Edinburgh.

Pearce, D.W. (1993a) *Economic Values and the Natural World*. Earthscan, London.

Pearce, D.W. (1993b) The economic value of externalities from electricity sources. Paper read to Green College, Oxford Seminar on Environment and British Energy Policy, April 1993.

Pearce, D.W. and Turner, R.K. (1990) *Economics and Natural Resources and the Environment*, Harvester Wheatsheaf, Hemel Hempstead.

Pearce, D.W., Bann, C. and Georgiou, S. (1992) *The Social Cost of Fuel Cycles*. HMSO, London.

Resources for the Future, and Oak Ridge National Laboratory (1992) Report to the US Department of Energy on the Social Costs of Energy. Resources for the Future, Washington DC.

3

French energy policy

R. Carle

3.1 INTRODUCTION

France is poor in energy resources and faced the oil crisis in 1973 in a particularly difficult position; 75% of energy was imported. This situation resulted in a strong energy conservation effort and programme to achieve a better equilibrium between the various energy sources. Today French energy independence is close to 50%.

The most efficient programme to achieve this objective was a large and rapid nuclear development launched in 1974. Now nuclear electricity represents 75% of the total electricity production. This programme was based on a standardized technology, large industrial investments and continuity. It included improved manufacturing quality, training of operation and maintenance personnel and particular emphasis on nuclear safety. The performances achieved are satisfactory. French investments include all necessary steps from uranium mining to waste disposal.

Besides its interest in terms of energy independence, nuclear energy has proved to be the most economical way of producing base-load electricity and environmentally the most benign form of electricity production. France, therefore, has a good record among industrialized countries regarding SO_2, NO_x and CO_2 emissions. Also France has developed electricity exports.

The French nuclear programme goes on. Its size is defined according to predictions of electricity consumption. It is now time to prepare for replacement of the present nuclear plants when they come to obsolescence around 2020. New models are under preparation as a continuation of the existing ones but with progress in terms of safety. This development should take place at a European level.

To begin with, it is important to recap on the main features of the energy situation of France. A few figures suffice to summarize this situation and explain the energy policy pursued by successive governments in France. France has considerable energy needs, amounting in 1992 to 221.5 million tonnes of oil equivalent, which contrasts sharply with France's limited natural resources. This translates into large dependence on imports and a weighty burden on France's balance of payments. The energy import bill rose from 17 billion francs (at current rates) in 1973 to 80 billion francs in 1992, 68 billion of which was for oil imports and 16.8 billion for natural gas. At the same time, energy exports generated 7.9 billion francs in revenues. In total, energy exports are equivalent to one twelfth of France's exports.

The ongoing objectives of French energy policy are more often than not contradictory: to ensure that the country has all the energy it needs, at the optimum price without depending excessively on imports. Depending on the circumstances, one or other of these requirements predominates in the policy pursued: using domestic coal reserves and hydropower resources in the immediate post-war period; widespread use of imported oil when it became inexpensive, for example during the 1960s; diversification of energy resources after the oil crises of 1973 and 1978.

For many years, Britain has always been able to rely as a source of prosperity on abundant coal reserves, and today on natural gas and oil from the North Sea.

France was not blessed with such good fortune: it has deep mines interspersed with faults, whose output has always been supplemented with imports. Nevertheless, France has extracted nearly 5 billion tonnes of coal, which is eight times less than Britain has been able to mine from its coal seams. But for the last 30 years, our coal has become too expensive to mine and since 1960, mining activities have been progressively wound down. From its peak of some 60 million tonnes in 1958, production declined to just 11 million tonnes in 1993. This decline is linked to the fall in consumption, which decreased to 30 million tonnes in 1992. Furthermore, very few of France's coal imports originate in Newcastle, in fact the majority – 80% – come mainly from the United States, Australia, South Africa, China and Colombia.

Nature was even less generous to France when it came to oil. What do the 3 million tonnes of oil amount to – which are obstinately extracted from our ground through determined efforts each year – in comparison with Britain's North Sea Oil reserves and compared, above all, with French needs, which currently stand at over 90 million tonnes?

As regards natural gas, although France does not have the equivalent of British offshore gas fields, it was fortunate enough to discover the Lacq deposits in the 1950s which harboured the equivalent of 250 billion cubic metres, which is far from negligible. Today, approaching depletion, these deposits yield barely 3 billion cubic metres of gas each year, which meets only one tenth of France's needs. French needs are covered by balancing supplies shipped from Algeria and from trans-European gas pipelines which flow from Siberian gas fields, and those from the Netherlands and the North Sea.

New energy sources should be mentioned in the hope that one day the breakthrough will be achieved which will make them available under competitive conditions. Today, France's principal energy source is electricity, which we call primary electricity: hydroelectricity and nuclear-generated electricity.

France is well endowed with potential hydropower resources. But despite this, hydroelectric power has never been able to supplant its thermal rival, contrary to the hopes of the pioneers of hydroplants in the Alps at the end of the nineteenth century. Exploited to the maximum of their economic and ecological capabilities, French hydro resources currently supply 70 TWR during average years of rainfall, which amounts to 23 million tonnes of oil equivalent.

It will come as no surprise to learn that much of France's electricity is nuclear generated; as much as 78% in 1993. Naturally, not all of France's electricity is produced using nuclear power. A three-pronged fuel policy is pursued based on uranium, water and coal, and it is important to retain this balance: a balance where nuclear power and 'run of river' hydropower facilities provide base-load supply; reservoir systems provide peak-load electricity; and coal-fired plants give semi-base-load supply.

Nevertheless, 78% nuclear-generated electricity is a remarkable achievement and a world first – excepting of course Lithuania, which has virtually no energy resources other than its Ignalina nuclear power plant. Furthermore, each year, France exceeds its own production record, illustrated by its performance over the last five years: 289 nuclear-generated TWh in 1989, 298 TWh in 1990, 315 TWh in 1991, 322 TWh in 1992 and 350 TWh in 1993.

This accomplishment is the result of a decision made in 1974 and maintained ever since with continuity and determination by successive governments. The success of France's nuclear policy is largely due to this perseverance, since the key to any nuclear programme is continuity, given the large time constants involved and the fact that a considerable number of years go by between the decision to build a power plant and its connection to the grid. Up to six years are needed if a site is available and if authorization has already been obtained, but this can stretch to eight years if the go-ahead has yet to be given. Continuity is also important because of the paralysing effect of calling into question a programme already underway, which proves so costly that the economic benefits of the installation are quickly eroded. Power companies in the United States are only too familiar with this type of setback. But continuity is important as well because it enables standardization, which in turn ensures a regular workload for equipment manufacturers, thereby both improving quality and reducing prices.

The decision taken in 1974 stemmed from the impact of the first oil crisis. Until that time, French energy policies were based on the massive use of oil, which was both plentiful and inexpensive. The tripling of crude oil prices, followed by the Yom Kippur War, drove home the extreme vulnerability of France *vis-à-vis* energy imports. Consequently, French policy was refocused on achieving independence from energy imports, by: renewing the priority given to coal (with limited results); geographic diversification of oil and gas supply sources; implementation of energy saving policies; and the launch of an extensive nuclear power programme aimed at replacing oil consumption with uranium consumption. The success of this policy is clear: independence from energy imports, which had sunk to 22%, has climbed to 50%, despite a parallel increase in energy consumption. From 183 million tonnes of oil equivalent in 1973, consumption grew to 222 Mtoe by 1992, despite the successful impact of energy saving measures, which have reduced consumption by some 30 Mtoe every year since their implementation. This substantial lessening of dependence is particularly impressive given that energy is such a basic commodity in our world today. For a country such as France, which is far from graced by natural energy resources, this is the reasonable limit we can hope to reach at a competitive cost.

Energy independence is mainly secured through primary electricity, which represents the equivalent of 85 Mtoe for hydropower and nuclear power together. France's electricity production is almost entirely 'home grown': all hydropower (for obvious reasons), one-third of fossil-fuel power (burning low-grade French coal products) and, above all, nuclear power, with all fuel cycle installations and power plants located on French soil. This is a crucial advantage, despite the fact that not as much importance is today placed on energy independence in a world where free trade has become the rule. The renewed abundance of oil and gas resources, together with the relaxation of prices, seems to be a natural consequence of free trade. But is this really the case and, more importantly, will it continue to be true? While not wishing to play Cassandra, it should not be forgotten that in the past 30 years, the world has experienced several periods of sudden, critical tension, which disproved all previous forecasts. Can we be sure that other such periods will not recur? EDF has prepared for this possibility by developing a high-performance production pool, of which the French are admittedly very proud.

There is much talk of France being 'isolated', or no longer 'believing' in its nuclear industry. Allusions to 'potential' incidents are increasingly heard, and the French nuclear power programme is said to create a gaping financial chasm, leaving in its wake vast indebtedness.

The author has no doubts whatsoever. He is a content and very proud nuclear operator, firmly committed to the nuclear industry. This belief is anchored in the four key characteristics of France's nuclear industry: it is operational, it is safe, it is environmentally friendly and it is economical.

3.2 NUCLEAR POWER IS OPERATIONAL

Generating over 75% of total electricity, nuclear power plants constitute the core of EDF's power supply system. The very low marginal cost of nuclear-generated electricity, which in practice is equivalent to the cost of the fuel alone, means that nuclear power plants are called on to generate electricity before any other means of production, except of course 'run of river' hydro-installations. The drought of recent years has led us to overlook the fact that in 1987 and 1988 nuclear power plants had been required numerous times outside their normal base-load operating periods. This demonstrated the ability of nuclear power plants to participate in load-follow programmes. This was necessary more than 2600 times in 1987, but only 1500 times in 1990, not because plants were unable to perform, but because they were needed to operate at full capacity for longer and more frequent periods during the course of the year.

With 56 units in commercial service, it is becoming difficult to plan all shutdowns for the summer period. However, every effort is made to avoid shutdowns for refuelling and maintenance during the peak-demand winter months. Ongoing efforts are made to reduce the duration of these shutdowns, which has resulted in a redefinition of refuelling policies in order to manage the operating programmes better. The fuel enrichment level has been increased to 3.7% for 900 MW reactors, which operate using the four-batch refuelling mode, with refuelling of uranium-235 carried out every year. In the 1300 MW reactors, for which three-batch refuelling with 4%-enriched uranium is used, the duration between shutdowns is to be extended to 18–20 months instead of 10–12. The performance of fuel elements is satisfactory, and this applies to both enriched uranium and mixed oxide (MOX) plutonium fuels. To date, five 900 MW units are loaded with MOX fuel. Plutonium recycling is also fully operational and the performance of mixed oxide fuel will progressively match the quality of enriched uranium.

The duration of shutdowns depends not only on refuelling, but also on maintenance and inspection operations, performed on equipment which is monitored very closely. The following example is linked to the use of the alloy Inconel 600, and concerns the behaviour of steam generators (in order to avoid tube rupture) and the systematic inspection of reactor vessel head penetrations in order to detect the appearance of cracks and to monitor their development. In both cases, specific monitoring tools have been defined. These include robotized inspection devices, readings from which are analysed, interpreted and recorded. Elements which are most affected by the identified faults are either repaired or replaced. Thus, the steam generators at one of the Dampierre unit have been replaced and the same operation has recently been completed at unit 5 of the Bugey plant. In all, 16 new steam generators have been ordered. As regards reactor vessel heads, which in the past attracted a great deal of media attention although installation safety itself was not in jeopardy, measures have been anticipated in the light of foreseeable developments. Repairs have been carried out on certain reactor vessel heads, and as from 1991 it was decided to replace the reactor vessel heads of the first six 900 MW units brought on line, and a total of nine reactor vessel heads were ordered in 1993. For all of these operations, from inspections and monitoring to repairs and replacement, methods were clearly defined, implemented and validated, benefiting from ongoing experience feedback.

Despite these anomalies, plant availability remains acceptable. Inevitably, availability declined in 1992 to slightly below the 72% target availability factor set when the nuclear power programme was first launched. But availability has since picked up, reaching 80% in 1993, a level which it is the intention to maintain.

3.3 NUCLEAR POWER IS SAFE

The safety of French nuclear power plants is fully guaranteed. Any fears of a serious accident are completely unfounded. By their very design, their solid construction and the presence of a containment building, France's reactors – like all reactors in the West – bear no resemblance to Chernobyl. While zero risk remains unattainable, the safety authorities believe that, in the highly unlikely event of a major accident at a French reactor, the maximum release of radioactive products would be only a few percentage points of that of the Chernobyl accident. This would imply, in the worst case scenario, either that inhabitants within the vicinity of the plant would be confined to their homes or, at the very worst, that they would need to be temporarily evacuated (to less than 5 km away). These measures alone would suffice to ensure their protection.

Apart from good design and good utilization, safety also depends on the quality of the nuclear 'culture' and on the quality of power plant operation. The safety culture is a veritable 'state of mind'; it must be evident in each and every operation or maintenance action. It is characterized by three underlying qualities: an enquiring attitude, a thorough yet cautious approach, and systematic, attentive communications. Plant personnel must be trained and made fully aware of their responsibilities. They require practical experience on simulators so that operating teams are prepared for any type of accident situation and are capable of performing diagnostics and implementing the appropriate measures. France is currently developing computerized maintenance procedures. But it must be emphasized that, far from 'superhumans', nuclear power needs operators of outstanding quality. Human resources (in the most fundamental sense of the term) are needed, supported of course by the latest-generation information systems, but which remain the most vital and best placed to deal with exceptional situations, far more than any conceivable automated process. It is out of the question for French plants to operate without the possibility of human intervention, or under circumstances where human presence would be considered unnecessary. Feedback consistently confirms the importance of human factors in every accident situation. Furthermore, special attention is paid to identifying, reporting and taking into account 'precursory incidents' to accidents.

This implies complete transparency, with regard to the management concerned, the safety authorities and the public. While some people tend to feel that something is always 'going on' in nuclear power, this is simply because everything that happens at the plant is reported. This also requires constant comparison of activities with those of foreign counterparts to ensure that safety culture does not remain insular. Furthermore, inspections such as OSART, conducted under the aegis of the IAEA, as well as the comparative safety reviews carried out with German or American reactors, have demonstrated that French installations implement equivalent safety practices. Finally, it implies open dialogue with the safety authorities, which remain entirely independent of the operator. This dialogue results in in-depth – and costly – inspections, such as those which EDF systematically conducts on reactor vessel heads and steam generators, and of course, where applicable, replacement of these equipment items as preventive measures. France ranks its safety-significant incidents according to the international nuclear event scale defined by the IAEA. The number of significant incidents is progressively declining, despite the fact that anomaly detection methods are increasingly sophisticated.

3.4 NUCLEAR POWER IS ENVIRONMENTALLY FRIENDLY

Every type of energy is confronted with obstacles in ecological terms. The nuclear power industry is relatively concentrated; it requires little of that rare commodity, space. In fact, in France today there are fewer power plants than there were before the nuclear power programme was launched.

Nuclear power avoids the emission of sulphur dioxide and nitrogen oxide; it also prevents releases of carbon dioxide, which are unavoidable each time combustion of oil, gas or coal takes place.

France has the advantage of holding the world record among the main developed countries: its CO_2 emissions are only 1.8 tonnes per inhabitant per year. Radioactive releases are also very low and strictly controlled. Indeed, they have much less impact on the environment than natural radiation levels. There is, of course, the heat factor, which is involved in any thermal cycle, but that aside, it can safely be said that nuclear power creates virtually no releases.

Many critics raise the question of nuclear waste. In France, there are storage facilities for low-level radioactive waste and efficient processing methods for high-level radioactive waste products have been developed.

All that remains to be done is to demonstrate the effectiveness of deep underground waste repositories, which means stripping bare what are largely psychological barriers.

In fact, nuclear waste, the volume of which is relatively small, is simply the precursor to all other waste produced by society, which should be stored in the same way as planned for nuclear waste products.

Is it possible to reduce the quantity and the half-lives of the most harmful waste? In theory, it is possible and studies are currently under way in this area. But is this really necessary? This problem involves the processing of all by-products from nuclear power, from plutonium to minor actinides. Work must continue in this field but this should not constitute a prerequisite to pursuing the nuclear power programme.

3.5 NUCLEAR POWER IS ECONOMICAL

First and foremost, the all too frequent slander which accuses EDF of being a subsidized company must be refuted: France's nuclear programme has been entirely financed by internal funding or through loans in the French and foreign markets.

During the last 25 years, the nuclear programme has absorbed 45% of EDF's investments and amounts to, including capitalized interest and commissioning costs, approximately 600 billion French francs at 1992 values. The State, EDF's share-holder, contributed only 50 billion francs, on which EDF pays interest. Furthermore, these contributions were terminated in 1981, and most of the nuclear programme has been funded from a combination of internal sources and borrowings. EDF's current debt of 170 billion francs is less than the company's sales, and financial charges are less than 12% of these sales. In constant francs, the sales price of the electricity has fallen by 20–30% since 1984, depending on the type of tariff in question. The 1993–1996 plan forecasts a 40 billion francs reduction in the debt and a further 1.25% cut in tariffs per year (in constant francs). Nuclear power has paved the way for these reductions and will continue to play this central role.

Operating expenses, which rose substantially between 1985 and 1990 largely due to the increase in the number of 10-year inspections, have since stabilized. Between 1990 and 1992, direct operating expenses totalled approximately 16 billion francs.

Taking into account these expenses, the price of the nuclear kilowatthour has stabilized at around 23 centimes, which is extremely favourable for base-load electricity despite the current very low costs of fossil fuels.

Nuclear power makes economic sense both for EDF and for France. Nuclear power generates the cheapest electricity in Europe, excluding a few very well situated hydroelectric installations favoured by their natural surroundings. The best illustration of the competitiveness of nuclear power is the sale by EDF in 1993 of 61.4 TWh to neighbouring countries at a price considered very attractive, even though it is higher than the price charged to French customers under the same terms. Exports could total 70 TWh in a few years. And it should be noted that countries buy not because of domestic shortages but because they find French prices extremely competitive.

Another important factor is that nuclear installations are paid for at the time of investment. This has two major effects. First, the price of nuclear power is sheltered from market fluctuations, hence the possibility of long-term contracts. This can mean, for example, a contract signed with the State which specifies a 1.25% reduction per year in tariffs, regardless of the rate of inflation. Second, significant cash flow is generated, which enables reductions in debt levels: 20 billion francs during the last contract and 40 billion francs in the next contract. The nuclear industry has a healthy debt level. The aim is to embark on installation renewal with a low debt level, accounting for approximately 5% of the price of the kilowatt hours.

Waste storage and decommissioning costs are allowed for and integrated into costs. Naturally, the actual level of these costs remains uncertain, but as their total final cost will be 2 to 3 centimes per kilowatthour, they will have only a marginal impact on competitiveness.

3.6 NUCLEAR POWER IN THE FUTURE

The French nuclear programme is being pursued at a steady pace of one new unit every two to three years in order to meet rising electricity demand, while awaiting the time, in 15 years or so, when all current plants will have to be renewed. The French nuclear power industry has a rich and active future ahead: for the 56 units in commercial service today, with an estimated operating life of 30 to 40 years for the four new units under construction and for the complete renewal of all facilities starting in about 2010.

The reactors of the future are being developed along the same lines as current installations, since the 50 years of operating experience accumulated to date are irreplaceable.

3.7 EUROPEAN UTILITIES REQUIREMENT (EUR) ACTION

In the late 1980s, EDF launched its REP 2000 programme, initiating a broad study on what should come after the N4 standard, around the turn of the century. These studies encompass experience feedback from operating PWRs and from the most recent R&D, exploratory work on so-called innovative ideas - with participation in the EPRI advanced reactor programme - and great attention paid to all international developments. One of the aims of that programme was to remain in the mainstream of world development without losing the benefits of the French programme and experience.

Several European utilities joined in this work in 1990 and by the end of 1991 they decided to work together to draft the European Utilities Requirements (EUR) document.

General (and generic) requirements will bear on topics such as size, grid requirements, fuel cycle, proven technology and simplification, safety objectives such as recommended by INSAG 3 and INSAG 5 and the IAEA Safety Fundamentals, and man–machine interface. They will also specify the methodology to be used for safety evaluation (particularly probabilistic methods and choice of input data) and for cost assessment. In order to contribute to international harmonization, the differences between the EUR and the EPRI requirements will be identified and kept to a minimum. There will be some differences, however, mainly due to the different regulatory contexts in the United States and in Europe, but also due to different fuel cycle policies and, to some extent, different grid requirements and siting considerations.

Specific requirements will evidently be adapted to each design and will be drafted accordingly. One of the most important issues is that of design and construction codes and standards. EUR has taken the decision that these will be specific and not generic, the acceptable codes and standards being, in principle, those proposed by the vendors (ASME for US vendors, Franco-German for EPR, etc.), with whatever additions may be necessary to meet regulatory requirements. This decision reflects the policy of European utilities not to restrict the market to European industry through technical barriers. It is not yet clear, however, which policies will be implemented by the European governments and licensing authorities: they might suggest that there be mutual recognition of models certified by the US NRC and by European licensing authorities in order to achieve reciprocity.

3.8 INTERNATIONAL CO-OPERATION

As already mentioned, European utilities view international co-operation as a major strategic objective, and not only where it means supporting the drive for enhanced safety in the nuclear plants of Eastern Europe.

During the past several years, there has been a growing effort to reconcile seemingly opposed safety concepts, involving inherent features, passive vs. active systems, automatic vs. human actions, small vs. large units, the need or not for emergency planning, etc. This effort has been driven, at least partly, by increased discussions between American, Japanese and European utilities, through co-operation agreements involving the EPRI advanced light-water programme and Japanese work on SPWR and the EUR programme. **It is of the utmost importance that this effort restores a main-**

stream consensus based upon common safety objectives and recognition that these common objectives can be met through different optimized concepts. If this effort were not successful, the public would not understand, and all the efforts to regain acceptance would have been lost.

3.9 THE FRANCO-GERMAN EPR

France and Germany teamed up to develop a common PWR design well adapted to European conditions and taking advantage of the wealth of experience built up over the past 20 years, and of the experience obtained from the most advanced reactors, the French N4 and German Konvoy. The EPR (European Pressurized Reactor), undertaken by NPI and the French and German utilities, will benefit from preliminary work undertaken by NPI (the 'common product'), EDF (REP 2000/N4+) and the German utilities' 'Planungsauftrag'. The French and German safety authorities have set up a joint *ad hoc* working group to define basic requirements, to evaluate the conceptual design and, later on, the basic design: this procedure should allow licensing of the design by both the French and the German authorities.

The harmonization phase leading to the conceptual design of EPR is well underway, with the three major objectives of visibly improving safety, of achieving greater simplicity than in existing plants, and of controlling the costs, while meeting the European Utility Requirements. This, of course, is easier said than done and requires, at each step of the design, always to favour a rational approach: the expected safety benefits must always be clearly and quantitatively identified (e.g. in probabilistic terms with regard to core damage and radioactive releases), the cost consequences evaluated and, most of all, the simplification potential analysed. Some preliminary conclusions may already be given.

EPR, at least the model designed for the French and German and most of the European market, is the nuclear island comprising a 4-loop 1400 MW unit, capable of load following and frequency control to meet the requirements of the European grid.

The reactor is designed to allow optimization of the fuel cycle to meet the needs of the first half of the 21st century: high burn-ups (up to 55 000–60 000 MWd/t) and plutonium recycling.

Severe accident mitigation is designed into the containment. The containment building is derived from the N4 design – a prestressed concrete inner vessel designed to withstand hydrogen deflagration and post-accident pressure build-up with adequate margins, an outer wall designed to withstand external aggressions, and a filtered ventilation system to cope with any leakage from the inner vessel. Corium retention and cooling capability will be provided below the reactor vessel, with a semi-passive system for ultimate heat removal from the containment.

Once the harmonization phase is over, and the French and the German safety authorities have reviewed the conceptual design, the basic design will be undertaken, leading to a PSAR in 1985 and a possible start of construction in France and Germany in 1998.

In parallel, R&D activities will be pursued by the German and the French research centres, particularly on the phenomenology of severe accidents and on fuel technology and reactor physics for high burn-ups.

3.10 THE NEED FOR FAST REACTORS

Fast reactors – or liquid metal reactors as the Americans call them today – used to be called breeder reactors because they have the capability of 'breeding' from uranium-238 more plutonium than they burn. In fact, breeders or not, fast reactors have the capability of multiplying by a large factor, compared to thermal reactors, the quantity of energy provided by natural uranium. The 'breeding' capability of the fast reactor is beneficial if the lack of plutonium is a limit to the penetration of fast reactors. When there is plutonium galore, it offers little interest.

This is obviously the case for the next decade(s), with reprocessing plants in Europe alone capable of extracting 20 or more tonnes of plutonium a year and with a limited capacity for plutonium recycling in light-water reactors (LWRs). When breeding will become necessary is today an open question, one of the main parameters being the total demand for nuclear energy. Meanwhile, fast reactors can best be used to demonstrate the technology and at the same time burn more plutonium than they produce (or at least not add to the surplus).

Fast reactors also have a potential for burning actinides, thanks to their high energy spectrum which allows fission of such isotopes as ^{237}Np and ^{241}Am. Although this may prove to be expensive, it is worth while testing as it could divide by as many as 100 the quantity of long-lived radioactive products to store.

3.11 THE PRESENT SITUATION IN EUROPE

The present situation of fast reactors in Western Europe is not very promising. PFR and Phénix are presently down for repairs, and the operating licence of Superphénix is suspended pending work on sodium fire mitigation. The owners of these reactors, whether R&D organizations or utilities, remain convinced, however, that fast reactors are a long-term insurance against a rise in the price of uranium and, as such, must remain an open option. Work is being carried on actively to bring all these units back on line as soon as technically and administratively possible. Superphénix will be used to demonstrate plutonium and actinide capability.

A future restart of Creys-Malville, with which the safety authorities are in favour and to which the government has agreed in principle, is hoped for. Restart will make it possible to produce electricity by optimizing the investments made in the installation and the fuel, and to acquire vital operating experience. It will also enable the development of methods for the destruction of plutonium used for military applications and of certain categories of long-lived radioactive waste.

3.12 THE FUTURE OF FAST REACTORS

Fast breeders could become a reality by the 21st century. British, French and German utilities, industry and R&D organizations have been carrying out and supporting the EFR project, a 1500 MW pool-type design derived from the previous work done in each country and particularly from the experience drawn from Creys-Malville, Italian, Spanish and Swiss utilities have shown an interest in EFR. Among the salient points related to safety is a thorough comparison of the respective merits of fully passive and of mixed passive/active systems for decay heat removal. A large cost reduction from that of Creys-Malville is expected, and the potential for further cost will be analysed, looking both at capital and fuel cycle costs.

Whatever decision is taken on EFR, European companies strongly support international co-operation. It is French belief that the existing agreements between Europe and Japan will allow sharing of work and of results on large fast reactors and on the corresponding fuel cycles.

3.13 CONCLUSION

France can be very proud of its contribution to the nuclear industry. It provides inexpensive electricity at a guaranteed price, ensuring supplies to energy-intensive industries. The nuclear industry is safe and poses no threat to public health or integrity, or to the environment, and its impact is substantially less than that of other power supply systems are being developed for the 21st century, which will almost certainly go beyond conventional fuels, in order to meet the ever increasing needs of a growing world population. The facts are known: from a population of 5 billion today, the world head count will grow to 8 billion in 2020, and current consumption of 8 Mtoe will rise to an estimated 12 to 15 Mtoe one generation from now.

As a key source of energy today, nuclear power is also the energy for the future. France is not alone in this belief, as illustrated by the efforts pursued to pave the way for future developments in conjunction with other power producers.

The West must of course work to avoid the recurrence of setbacks, illustrated by the vast problems facing the nuclear sector in Eastern Europe. Although vital to their needs, these installations were poorly constructed, with insufficient safety standards. They must be improved and upgraded, and it is in our common interest to assist our partners in the East in raising safety standards.

In a world where communications play an increasingly important role, a consensus must be fostered regarding nuclear power. Failing this, if we are unable to continue to optimize this precious energy source, and if we fail to overcome the reserves harboured by some members of the public, future generations will inherit a legacy of energy shortages and perhaps also of conflict.

Nuclear power is one of the cornerstones of French energy policy. Energy consumption is not expected to rise significantly in the near future. Even if coal maintains its 'honourable' position and if natural gas increases in relative importance, nuclear power and oil combined will continue to meet nearly three quarters of French needs. The answer lies not in classifying energy types into 'good' and 'bad', but in solving, through a reasonable approach, the true challenge already confronting the southern regions of the world: strong population growth. The author is confident that the challenge can be met.

4

Fossil power generation and sustainable development

G.T. Preston

4.1 INTRODUCTION

Electricity offers a sustainable solution to the global issues of population and poverty, yet half the world's people do not have access to commercial electricity. Rapid advances in utilization of fossil fuels for electric power generation – by state-of-the-art pulverized coal firing, fluidized bed combustion, integrated coal gasification, combustion turbines and fuel cells – have long-term implications for developing and transitional economies as well as developed countries. In the developed economies, economic and environmental pressures will sustain the 'rush to gas' for the near term – to such a degree that in some countries sunk investments in existing coal-fired plants are at risk. In contrast, for the near term the fastest growing developing countries with indigenous coal resources will choose medium and large central station generation – with the greatest priority on low cost and high efficiency and less emphasis (for now) on environmental considerations. Continued fossil power technology advancements, and technology transfer to accelerate their implementation, hold promise to help meet global development needs in the near term while laying the foundation for sustainability over the long term.

4.2 THE GLOBAL PROMISE OF ELECTRICITY

The theme of the meeting – Energy Technology for the Third Millennium – appropriately implied a world-wide perspective for the discussions. Accordingly this chapter is focused on the future for fossil power generation technologies in the context of global sustainable development.

Population and poverty are the paramount long-term issues faced by the global village. The enduring solutions are global economic development, education and electrification. The means to achieve these solutions are technology innovation and application.

4.2.1 Energy efficiency and electrification

An advanced stage of human development and quality of life does not inevitably require or imply high energy consumption. In fact, the past three decades of experience in the United States, Japan and the highly developed European countries show that increasing sophistication in industrial progress leads to decreased intensity of energy consumption (Figure 4.1). For example, for the past several years growth in US energy consumption has been less rapid than growth in US gross domestic product (GDP). Specifically, energy consumption per unit of GDP decreases as electricity accounts for a greater fraction of ultimate energy use.

The implication of this relationship for energy policy planning is profound, given that in many of the developing economies the growth rates of both electricity consumption and GDP are two or more times average world growth rates. The message is that increased electrification is a key mechanism for accomplishing economic development while using energy resources effectively, equitably and environmentally responsibly.

Two billion people in developing countries do not have access to commercial electricity. The developing countries have 76% of the world's population and account for only 22% of world electricity consumption. World electricity demand increased during the 1980s at 3.5% annually; electricity demand in the developing countries has grown at 8.3% annually

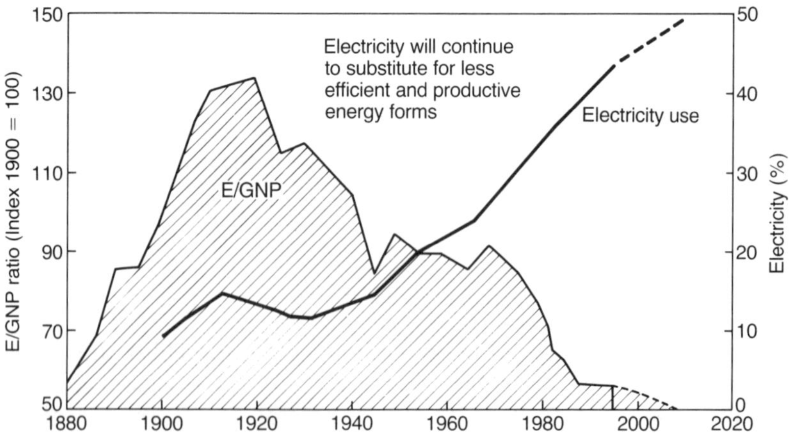

Figure 4.1 Energy consumption/GNP vs. electricity use. (Source: *Energy Outlook*.)

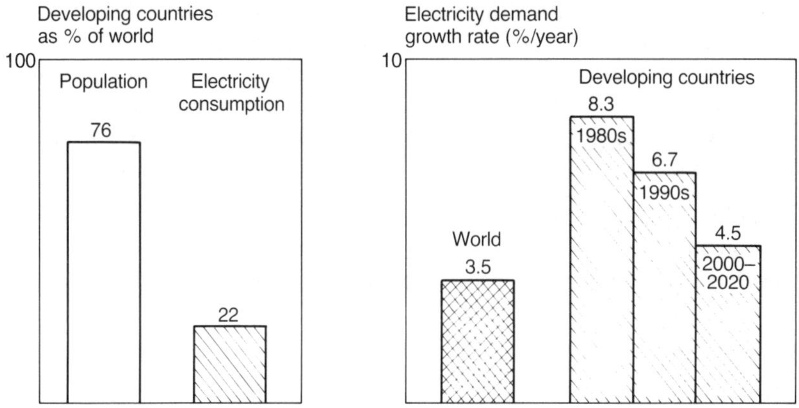

Figure 4.2 Projections of energy consumption by today's developing countries.

over the same period. In the 1990s the projected growth of electricity demand in the developing countries is 6.7%, and during 2000–2020, 4.5% (Figure 4.2). Moderate projections of population and economic growth bring us to the expectation that by mid-21st century 60% of global electric consumption will be in the current developing countries.

Following is a brief overview of the major fossil resource based electric generating technologies that will be important to world-wide sustainable development in the next several decades.

4.3 PULVERIZED COAL

Coal is the most abundant fossil fuel, accounting for 80% of the world's fossil reserves. The geographical diversity of its resource locations, low extraction costs, low transportation costs and production capacity in excess of world requirements keep the delivered price of coal at the low end of the fossil fuel spectrum. Most projections show that world-wide coal consumption will at least double in the next 30 years.

The goal of a pulverized coal (PC) power plant with near zero wastes and thermal efficiency approaching 50% is technically possible within the first decade of the 21st century. PC-fired electricity generation will remain the technology choice for many US utilities and should be attractive in transitional and developing economies for several reasons:

1. The PC power plant is proven technology for small or large capacity increments up to 1300 MW using a single-shaft steam turbine.
2. The supercritical steam cycle technology used in new power plants being installed now has the capability for over 40% thermal efficiency even with full environmental controls. By early next decade a PC-fired

combined cycle with topping turbine could achieve 47% thermal efficiency in a plant with full environmental controls.

3. Current technology allows reliable and effective control of sulphur and nitrogen oxides. The experience base in the United States and elsewhere is growing steadily. For example, by the year 2000, 60 GW of US PC-fired generating capacity will be equipped with SO_2 scrubbers.

Technology is available from world-wide suppliers to build new state-of-the-art power plants (SOAPP) incorporating the following features, reflecting the results of decades of operating experience and development progress by EPRI, US and international utility organizations, equipment manufacturers and engineering firms.

1. 'Ultra supercritical' steam conditions of 31 MPa (4500 psig), 870 K (600 °C, 1100 °F), double reheat.
2. All-ferritic alloy boiler and turbine materials.
3. On/off cycling capability; operable as low as 10% load; 40% thermal efficiency down to 50% load.
4. Modular (replicated and highly shop-fabricated) construction; 24-month construction time from first concrete pour.

4.3.1 Efficiency improvement and environmental performance

As pointed out earlier, electrification is a path to improving overall primary energy utilization. Recent advances in coal-fired power technology offer opportunities for improving the efficiency of the conversion step from primary energy to electric energy.

1. Most new coal-fired plants built in the industrially developed countries will employ supercritical or ultra supercritical steam cycles and advanced plant control systems to optimize thermal efficiency at any required load.
2. Thermal efficiency can be improved by at least 3–5% in most existing large fossil power plants by installing heat rate and monitoring instrumentation and implementing formal heat rate improvement programmes with rapid economic payback in recovered generating capacity and/or fuel saved. Such thermal efficiency improvements naturally result in 3–5% or greater reduction in particulate, SO_2 and NO_x emissions for the same electric output. Thermal efficiency advances reported at recent EPRI conferences and other international forums include: (i) computer software to model boiler cleanliness, heat balance and gas duct dynamics; (ii) new instruments and systems for testing thermal performance, monitoring heat rate, measuring circulating water flow and measuring low-pressure steam turbine efficiency; and (iii) auxiliary equipment improvements such as condensing heat exchangers, macro biofouling

control techniques, condenser air leakage control, air heater monitoring and feedwater heater control.

4.3.2 Improving the use of existing plants

Where an installed base of coal-fired electric generating capacity already exists, improving operation and maintenance along with implementing moderate-cost efficiency and environmental upgrades can be the most cost-effective near-term as well as long-term approach to generating capacity planning. This means that for existing plants also, recent technology advances offer numerous opportunities. Some examples are described briefly below.

Identify and permanently solve failure mechanisms

Many PC-fired plant shutdowns are caused by component failures whose root causes are now understood and preventable through relatively inexpensive operating changes, plant monitoring or component modifications. Examples of the current state of knowledge are EPRI's Boiler Tube Failure Reduction guidelines, training curriculum, and the computerized Boiler Maintenance Workstation.

Restore or improve thermal efficiency through performance monitoring

Performance monitoring is the on-line detection of deteriorating thermal efficiency and its root causes. EPRI and the Potomac Electric Power Co. (PEPCO) have been developing improved sensors and instrumentation systems since 1982 at PEPCO's Morgantown station. The computerized Performance Monitoring Workstation compares measured and target performance, looks for effects of operating changes on plant performance and pinpoints component degradation. Several other US utilities have installed for are planning to install the Performance Monitoring Workstation at existing power plants.

Implement diagnostic monitoring and predictive maintenance

Present technology in continuous and intermittent diagnostic monitoring allows a maintenance philosophy called predictive maintenance that is based on knowing the condition of each plant component - rather than on a maintenance calendar (preventive maintenance) that can sometimes result in costly and unnecessary replacement of components. Predictive maintenance is being demonstrated by EPRI and Philadelphia Electric Co. at the Eddystone plant in Philadelphia.

Retrofit modern control systems

Most PC plants operating today were designed with analogue or pneumatic control systems that are 15–20 years old or more. Modern digital distributed control systems offer the potential for intelligent automation and integration with plant monitoring data.

Train operators with low-cost plant simulators

The power of the modern microprocessor allows desktop workstation simulators to be designed and built for a fraction of the cost of the few full replica simulators that were built in the past for fossil power plants. Such simulators can be used to test operational changes and optimize startup and shutdown sequences as well as for new and refresher training of plant operators.

Upgrade deteriorated or obsolete plant hardware components

Examples of modern components commercially available are:

1. retrofit burners for low NO_x and low load operation;
2. super 9Cr steel alloy for longer life, higher-temperature thick-wall components such as headers and steam lines;
3. improved water chemistry instrumentation;
4. titanium alloy low-pressure turbine blading;
5. super clean steel alloys for low pressure turbine rotors;
6. erosion-resistant coatings for high pressure turbine blades and valves;
7. improved generator stator insulation;
8. inherently stable feed pump designs;
9. heat pipe air heaters;
10. adjustable speed drives.

US design, engineering and operating experience is already available for many of these upgrades.

4.4 COAL UPGRADING

Coal upgrading is a strategy for using indigenous fuels economically. While serious development and demonstration efforts will continue to remove organically bound sulphur from coal, for most fuel choice situations in the near term, coal upgrading will mean physical cleaning to remove inorganic (ash forming) constituents and pyritic sulphur. There is an extensive base of US experience with coal cleaning in more than 400 operating coal cleaning plants, most of them at the coal mine. The economic incentives and effectiveness measures for coal cleaning are shown below.

1. Reduce transportation volume and weight with minimal loss of the fuel's energy content.
2. Reduce the impact of ash constituents on the power plant's coal handling equipment, pulverizers (erosion), boiler tubes (slagging, corrosion–erosion), particulate control and solid waste handling system - all of which affect operating and maintenance costs and thermal efficiency.
3. Reduce sulphur content to avoid or reduce the cost of SO_2 control and solid waste disposal systems.
4. Improve the homogeneity of the fuel as delivered to the plant's boiler.

Even for lower rank (lower value) sub-bituminous coals and lignites, for specific situations a comprehensive analysis of the true fuel cost, i.e. including impacts on the plant, may show a payoff for coal cleaning that could not be justified based only on delivered fuel price per energy content. EPRI's Coal Quality Impact Model, available commercially, and the more comprehensive Coal Quality Expert being developed by CQ, Inc., under EPRI and US Department of Energy funding, are tools for evaluating the economic tradeoffs in coal selection and coal upgrading decisions.

4.5 FLUIDIZED BED COMBUSTION

Fluidized bed combustion (FBC) has several attractive features as a coal utilization option:

1. It offers broad flexibility of coal feedstocks, compared with pulverized coal or gasification plants.
2. It has inherent SO_2 and NO_x reduction - SO_2 by *in situ* absorption by limestone fed into the fluid bed, NO_x due to significantly lower combustion temperatures relative to a PC boiler.
3. It is cost-competitive with PC-fired plants, especially at the smaller sizes that may be most practical for bringing electricity generating capacity to medium-size growing urban load centres.
4. There is an extensive experience base, especially in the smaller sizes found in several decades of US industrial application.

4.5.1 AFBC

There are over 150 atmospheric FBC (AFBC) units operating for power generation in the United States and 10 vendors of FBC boilers, four of whom offer units as large as 250 MW.

Large-scale bubbling bed (most of the solids stay in the bed) AFBC units are operating successfully at Tennessee Valley Authority's Shawnee station in Kentucky (160 MW) and Northern States Power's Black Dog plant in Minnesota (133 MW). The focus of AFBC in the United States is now on circulating fluid beds (CFBC - most of the solids continually circulate and are returned to the bed). The repowering project at Nucla station in Colorado (currently Tri State Generating & Transmission Assn. Inc.) successfully demonstrated CFBC technology at 110 MW scale. Texas New Mexico Power Co. built the largest CFBC units in the world - 2 × 150 MW lignite-fired, with reheat - and started them up in 1990. Commercial plants in planning have reached the 250 MW size in the United States and 350 MW in Japan.

EPRI evaluated early operating experience on four second-generation AFBC boilers (including the two Texas New Mexico units) to compare their availability with 27 PC boilers, 150–800 MW in size, all with flue gas SO_2 scrubbing systems. The average availability of the four AFBC units was only two percentage points less than the 27 PC units (86% versus 88%) - indicative of an advanced stage along the learning curve for the first large-scale applications of this technology.

Most US AFBC units today are being installed in the 50–100 MW size range by independent power producers. Fuel feedstocks include not only coal and lignite, but also petroleum coke, anthracite culm, forest wastes and tyres.

4.5.2 PFBC

Compared with AFBC, pressurized fluidized bed combustion (PFBC) has the advantages of smaller size and thus lower capital cost, lower fluidization velocity (less erosion), added fuel flexibility, more opportunity for shop-fabricated modular construction and the potential for thermal efficiency over 40%.

The operating experience base is more limited for PFBC. Large-scale US experience resides in American Electric Power Co.'s Tidd station 70 MW repowering demonstration under Round 1 of the US DOE's Clean Coal Technology programme. Other operating experience is being acquired from 2 × 80 MW bubbling bed PFBC units at Stockholm Energi's Vartan station in Sweden and from Endesa's Escatron 80 MW unit in Spain. The Stockholm units are providing district heating as well as power. The Escatron unit fires a lower-rank 'black lignite' with high ash, high sulphur and high moisture content. All four of these units have demonstrated environmental performances better than expected.

Other PFBC projects include Electric Power Development Co.'s 70 MW demonstration in startup at Wakamatsu, Japan; Midwest Power's planned 70 MW circulating PFBC repowering under Round 3 of the DOE Clean Coal Technology programme; and American Electric Power's deferred 320 MW unit under Round 2 of the CCT programme.

Recent EPRI studies suggest that the cost of electricity from a bubbling PFBC plant burning high-sulphur coal is competitive with other advanced fossil technologies and may have significantly lower cost than the other options when low-sulphur coal is burned.

Circulating PFBC is in the initial development stages, as is advanced PFBC, described briefly later.

4.6 ADVANCED COAL TECHNOLOGIES

The technologies reviewed above are based on the Rankine steam thermodynamic cycle and (except for SO_2 and NO_x control in FBC) rely on post-combustion clean-up of environmental pollutants. Owing to their extensive experience base in the developed countries, these technologies are expected to meet current and near-future economic efficiency and environmental performance requirements, especially in transitional and developing economies. However, a longer economic and environmental sustainability perspective requires attention to higher efficiency, inherently cleaner technologies. Moreover, some of these advanced technologies - notably coal gasification in a combined cycle configuration - are already competitive with state-of-the-art PC-fired electric generation in some applications.

The leading near-term advanced coal technology is integrated coal gasification combined cycle (IGCC) of which several plants are in demonstration. In the future, significant efficiency improvements can be expected from combining coal gasification and pressurized fluidized bed combustion technology - termed advanced pressurized fluidized bed combustion or APFBC; and ultimately the integration of coal gasification and molten carbonate fuel cells - 'integrated gasification/fuel cell' (IGFC). APFBC is being pursued by several developers and will be demonstrated at pilot scale in a joint project of the US DOE, the Southern Company and EPRI in Alabama. As for IGFC, the first fuel cell operation on coal gas occurred in September 1993 in an EPRI-sponsored test at Destec's Plaquemine, Louisiana gasifier. The challenge for these latter technologies - which have the potential for thermal efficiencies approaching 60% - is to reduce their complexity and capital costs to keep them 'in the running' in the competition with more conventional technologies that are also continuing to evolve.

The US Department of Energy's Clean Coal Technology Program, initiated in 1986, has been essential to the demonstration, and indirectly to the continued development, of advanced coal technology in the United States during an era when the availability of natural gas and the relatively low capital cost and risk of combustion turbine capacity additions has temporarily dampened enthusiasm for new coal-based generation. The DOE co-sponsored demonstrations include not only the application of advanced high performance emission control systems to existing PC plants, but also first-of-a-kind advanced power generation systems such as PFBC and IGCC. DOE has also built links from advanced coal-based development programme into its Advanced Turbine Systems programme aimed at commercialization of 60% thermal-efficient (lower heating value) gas turbine cycles.

4.6.1 Coal gasification

Coal gasification technology has a long history in providing the industrialized world with fuel gas, illuminating gas, chemicals, liquid fuels and transportation fuels. The application of coal gasification to central station electric power generation is more recent. Large-scale demonstrations are summarized in Table 4.1. These plants fuel combustion turbines operating at combustor outlet temperatures about 1370 K (1100 °C, 2000 °F) and have overall thermal efficiencies about 41% based on higher heating value. Acid gases are removed before the gas turbine using cold gas clean-up technology already commercial in the petrochemical industry.

Higher thermal efficiency (about 2% higher) will be achievable by using hot gas clean-up technology and the currently available 1560 K (1290 °C, 2350 °F) combustion turbines. Planned large-scale demonstrations of this generation of IGCC are summarized in Table 4.2.

Table 4.1 Large-scale coal gasification projects

Name/location	*Process*	*Size (MW)*	*Operation*
Cool Water Daggett, CA	Texaco	100	1984–1989
Destec Plaquemine, LA	Destec	160	1987–
SEP Buggenum, The Netherlands	Shell	253	1993–

Table 4.2 Planned IGCC plants

Name/location	*Process*	*Size (MW)*	*Operation*
RWE Germany	HT Winkler	300	1996
Endesa Spain	Prenflo	330	1996
Tampa Electric Polk County, FL	Texaco	260	1996
PSI Energy Wabash River, IN	Destec	268	1995
MITI Japan	not selected	370	2000
Sierra Pacific Piñon Pine (Reno), NV	KRW	95	1996

4.7 COMBUSTION TURBINES

As the electricity business in the United States becomes more competitive, and other developed nations' state-owned electric utilities move towards privatisation, new economic criteria are favouring gas-based generation and distributed supply options.

For most US situations the economically and environmentally preferred approach to adding generating capacity is natural-gas-fired combustion turbine simple and combined cycle units. With continuing low prices of natural gas in the United States this strategy can offer the least-cost electricity available for a new plant. Even utility companies who plan to add SO_2 scrubbers or switch to low-sulphur coal for their existing coal-fired plants are considering combustion turbines for capacity growth because of their low capital cost, low emissions, short lead time for installation and adaptability for repowering ageing coal-fired plants. Figure 4.3 shows that in the United States 97 GW or 86% of new capacity in the next 10 years is expected to be combustion turbine or combustion turbine combined cycle; world-wide, combustion turbines will account for a total of 288 GW or 47% of new capacity.

The efficiency and performance of advanced heavy-frame combustion turbines are increasing as equipment suppliers continue to raise firing temperatures and unit capacities. The current state of the art is manifested in the 'F-class' machines now offered by all four heavy-frame combustion turbine suppliers, firing at 1260 °C or higher combustor outlet temperature. The next generation of heavy-frame machines will have firing temperatures of 1370–1430 °C and combined cycle thermal efficiencies approaching 60%, and will be commercially available by the year 2000.

Performance advances in the high-thrust fan jets that power large jet airliners are the basis of growing interest in aeroderivative turbines. These

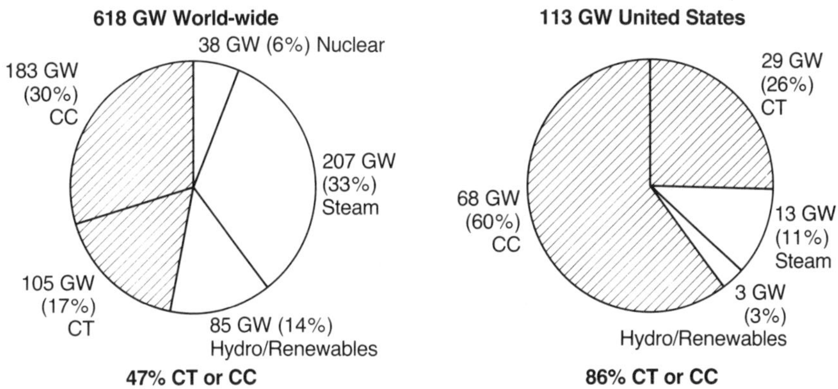

Figure 4.3 World and US capacity additions, 1993–2002. (Source: General Electric.)

are smaller and more efficient (in simple cycle) than the heavy-frame machines. They cost more per kilowatt capacity than heavy-frame turbines but have the advantage of a prepackaged shippable modular configuration that can generate electricity within hours of being delivered to a site.

The Collaborative Advanced Gas Turbine (CAGT) project is an industrial consortium managed by EPRI and including the world's major aircraft engine manufacturers and American, British, Canadian, Danish, Dutch and French utilities. CAGT is working to accelerate the development of advanced high efficiency aeroderivative gas turbines incorporating the latest fan jet engine technology. An early intermediate deliverable from this effort will be the intercooled aeroderivative engine, which manufacturers expect to offer by 1997–1998 in 80–100 MW capacities with 47% thermal efficiency.

Figure 4.4 shows the potential for progress in the performance of aeroderivatives compared with heavy-frame gas turbines in simple cycle application. The intercooled aeroderivative gas turbine in simple cycle will offer a thermal efficiency almost as good as a current heavy-frame combustion turbine in combined cycle, at a cost only a little higher than today's heavy-frame simple cycle machines.

Advanced gas turbine cycles such as the steam injected gas turbine, chemically recuperated intercooled steam injected gas turbine and humid air turbine will further increase efficiencies and lower the cost of electricity; moreover, these advanced cycles have the potential to bring even closer together the performance and cost of aeroderivatives vs. heavy frame turbines.

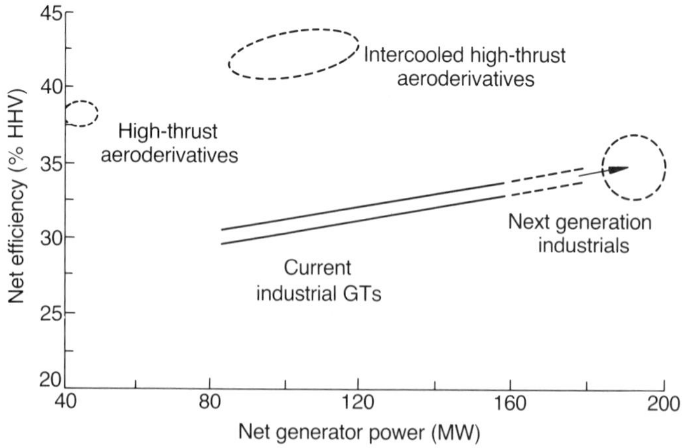

Figure 4.4 Simple cycle GT options for 1997.

4.8 FUEL CELLS

Unlike conventional combustion-based fossil fuel generating technologies, fuel cells convert fuel energy to electricity through a flameless oxidation reaction, as in a battery. Phosphoric acid fuel cells (PAFCs) operating at about 200 °C with a thermal efficiency of 35–45% were first operated for electricity generation by Tokyo Electric Power Company in 1983. For utility scale application US attention has been focused on molten carbonate fuel cells (MCFCs), which operate at 650 °C with a thermal efficiency of 54–60%[1] and can be smaller than PAFC plants owing to their internal reforming of the fuel gas within the cell stack. Figure 4.5 shows a development and commercialization timeline for both gas fuelled and coal-gas fuelled MCFCs.

A 2 MW demonstration plant consisting of 8 × 120 KW MCFC stacks is in final design and fabrication by a consortium headed by the City of Santa Clara, California, and including EPRI and the US Department of Energy. The demonstrated technology would cost about US$1500/kW in commercial production, which would be competitive in many US applications. However, by design and manufacturing improvements in the stack modules and the balance of plant, the costs are expected to decrease to US$1000–1100/kW.

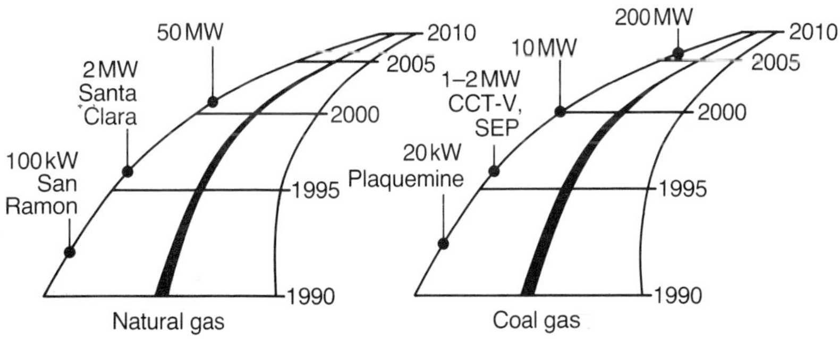

Figure 4.5 Fuel cell timeline for MCFC plant commercialization.

[1]Thermal efficiencies in this chapter are based on lower heating value of the fuel unless higher heating value is stated. Thermal efficiency based on higher heating value is approximately 2–4% lower for coal, 3–6% lower for gas, than efficiency based on lower heating value.

Solid oxide fuel cells (SOFCs) operate at about 1000 °C but are expected to have a lower thermal efficiency. Development currently is at the 100 kW pilot plant stage.

4.9 DISTRIBUTED SUPPLY

The increasingly competitive electric business environment in the United States and other developed countries, and the growing economic regulatory and public acceptance issues associated with transmission and distribution, are driving the rapid emergence of a new business strategy - distributed supply or distributed resources - referring to any modular generating or storage technology that can be used to improve service to the electricity consumer and optimize the use of utility corporate assets. Figure 4.6 illustrates commercially available distributed resource technologies.

Figure 4.7 shows the applications and size range for which distributed generation technologies are most suitable, along with magnitudes of capital cost.

In the United States over half of the embedded investment in the regulated utility industry's capability to provide electricity is now in the transmission and distribution system. Historically, distribution systems have been designed to be able to deliver peak electric demand all the time. When the transmission and distribution (T&D) delivery system reaches its capacity owing to regional load growth or new demand resulting from regulation-mandated access to outside electricity producers, in a growing number of situations it makes sense to look for alternative ways to supply the customer for the most costly hours in the year. Figure 4.8, based on a real-world example, shows that during 70 hours of the year it would cost over US$1/kWh (i.e. about 10 times typical retail electricity prices) to deliver electricity to the customer in this particular distribution area if T&D

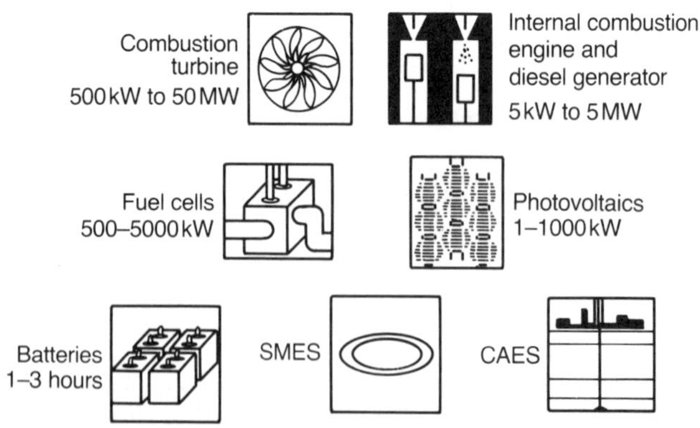

Figure 4.6 Distributed generation and energy storage technologies.

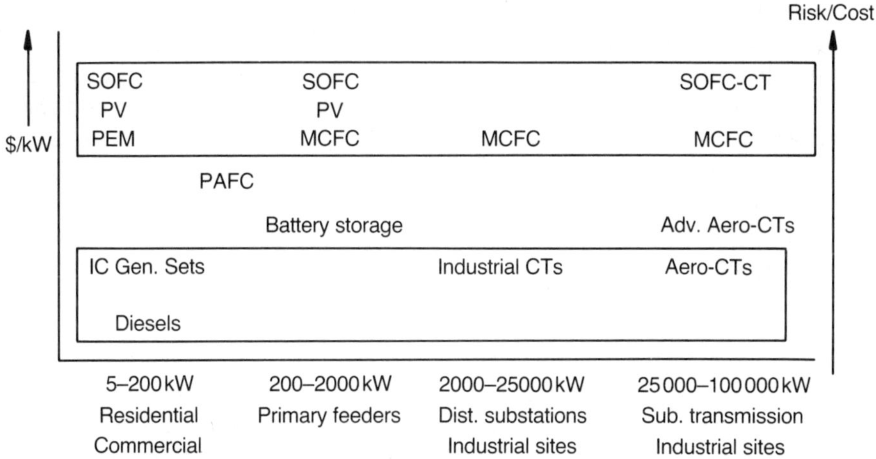

Figure 4.7 Distributed generation technologies – applications and sizes.

capacity were increased to meet the new demand. The area between the cost vs. hours curve and the battery cost line is a measure of the benefit of using a distributed resource, e.g. batteries, as an alternative way to provide electricity during those costly hours.

The distributed resources concept is focusing new attention on small-scale fossil generating technologies – notably fuel cells and small combustion turbines – and altering many utilities' traditional goal of 'be the low-cost producer' to 'be the provider of choice'.

4.10 PERSPECTIVES

4.10.1 Electricity business environment

The economic, regulatory and social environment in which industrialized world electric power generation businesses operate is changing rapidly in several ways.

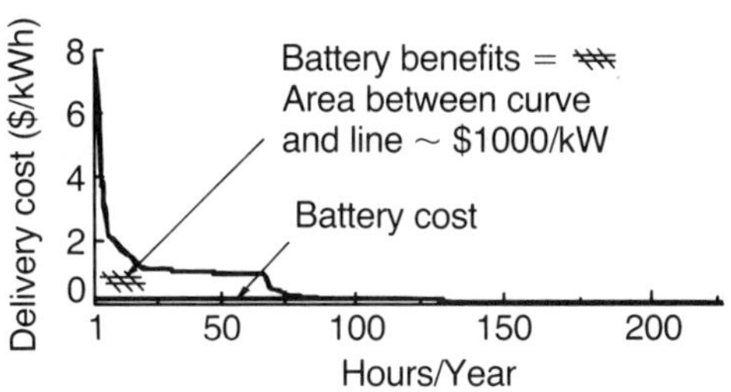

Figure 4.8 Distributed batteries avoid new transmission and distribution investments.

1. The electric customer is more sophisticated and more demanding. Customers want more influence on the business direction of their electricity supplier, and they expect more breadth of choice in the services offered.
2. The composition of the industry – the number of participants and the character of the participants – is changing. Government power authorities are being privatised – nowhere more visibly or dramatically than in the United Kingdom, but in several other countries also. Traditional vertically integrated investor-owned utilities are considering a variety of dis-integration or unbundling options. Non-utility generators have entered the power generation business. Mergers, acquisitions, joint ventures and alliances are occurring as utilities look for synergies to cut their fixed costs, stay competitive and open up new markets.
3. The regulatory framework is changing in the United States and other industrialized countries. In the United States in the 1980s a few electric utilities experimented with diversification and disintegration before the regulatory climate was ready. In contrast, the rules are changing so fast now that many companies are running to catch a train that is already moving out of the station. Examples of regulatory changes in the United States are the 1992 National Energy Policy Act with increased transmission access; evolving environmental regulation with increasing emphasis on pollutant prevention and externality-based cost incentives; and integrated resource planning.

4.10.2 Fossil capacity in the United States

In recent years coal as an electricity source has accounted for over half of all US electric generation in kilowatt hours. Coal's share of electricity generation will decline slowly, but even by 2010, US Energy Information Administration projections forecast that 48–52% of US generation will be coal based. However, in most regions of the United States there is still an oversupply of electric capacity – the capacity needs are primarily to satisfy growing peak demand – so that little base-load generating capacity will be added in the 1990s.

This prognosis and the increasingly deregulated competitive business environment mean that most US attention to coal-fired electric generation is:

1. by regulated utilities who plan to continue operating their existing coal-fired generation as long as they can do so competitively;
2. by regulated utilities who want to preserve their option to build new (advanced technology) coal-based capacity after the year 2000;
3. by independent power producers aiming at small to medium sized, low-cost, no-frills FBC or pulverized coal plants in least-cost open-bid situations;

4. by independent power producers or non-regulated utility affiliates looking towards international power markets.

Economic and environmental pressures in the United States and other developed countries will sustain the 'rush to gas' by independent power developers and other utility competitors for the near term – up to about the year 2000 – long enough for some existing coal-fired plants to be at risk of being uneconomic and written down on their owners' books.

4.10.3 Distributed coal-fired resources

Rapid build-up of central station generating capacity is not the only technology need in transitional and developing economies. Away from densely populated urban areas and for transmission and distribution system stability and flexibility, coal-fired units in the 100–350 MW range may be preferred. AFBC and PFBC are likely to compete favourably with PC-fired plants in such situations.

4.10.4 Gas and gasification

US and world reserves of gas seem plentiful, but availability and deliverability at economic prices create an issue that suppliers and consumers see from two different viewpoints. Gas enjoys an advantage in the global climate debate because of its lower CO_2 emitted per unit of electricity. The issue of leakage of methane (a potent greenhouse gas) to the atmosphere in the distribution system needs to be quantified, although the amounts appear to be small, at least in the United States.

Recognizing the world's plentiful and geographically well-distributed resources of coal and the economic growth pressures in developing countries, IGCC must be included in any near-term oriented discussion of 'sustainable' electricity technologies. Its efficiency and environmental performance surpass conventional Rankine-cycle pulverized coal generation, and it is already cost-competitive with pulverized coal generation in many applications. Moreover, the prospect of further cost reductions and efficiency improvements effectively puts a cap on the market price of natural gas and will reduce the advantage of gas as a low CO_2 producer.

4.10.5 Capital requirements

Significant capital investment is a characteristic of the fossil generating technologies most suitable for accelerating economic development in the transitional (e.g. Eastern Europe and some Pacific Rim) and developing (China, India, Indonesia, Philippines) economies. This suggests that development of financing capabilities and financial infrastructure is likely to be a pacing issue for near-term growth of electrification in these countries.

4.10.6 What is 'sustainable'?

The theme, 'sustainable development', raises interesting issues - most of them not technical - when the electricity growth needs of the world's developing economies are considered. Lack of access to commercial electricity is a major barrier to sustainable development in these countries. How can they invest - and how can we help them invest - in electrification in a way which we would regard as 'sustainable'?

Fossil steam generating plants produce about 70% of the electricity in the developing countries, and hydroelectric 29%. There is plenty of potential for hydro development, but this implies access to capital, a developed technology infrastructure and a transmission and distribution system. This means that the developing countries that now have coal-fired electric generation and indigenous or accessible coal resources are likely to expand their electric capacity primarily with coal. For the near term, the most practical contribution of the developed economies to the sustainable electric development of the developing countries may be to transfer recent evolutionary improvements in technologies for which there is already some established experience base.

4.11 CONCLUSIONS

1. Global electrification is essential to sustainable development and is part of the long-term solution to population and poverty.
2. The future for electric power generation from fossil resources will be heavily influenced by thermal efficiency and environmental considerations.
3. Although most added US electric generating capacity in the 1990s will be gas combustion turbines, half of all US electric generation will continue to be coal based for the next 25 years, and the fastest growing developing countries will be relying initially on coal for electricity growth.
4. Distributed supply concepts not only represent an important strategy for utility operations in the more competitive business environment in many developed countries but might also be applied in less-populated areas of developing countries to get an early start on electrification.
5. High efficiency gas turbine cycles in development now will benefit coal gasification technology costs and performance. The highest-level coal-based electric generation - ranked against sustainability criteria - will be coal gasification integrated with fuel cells.
6. Technology advances in fossil power generation can contribute to

sustainable development most effectively if there is a common recognition that 'sustainable development' is a concept whose definition is sensitive to time-and-place context. In the next 25 years sustainable development in the world's transitional and developing economies will mean a heavy reliance on fossil power generating technologies that can be implemented at relatively low cost and have an extensive knowledge base already established.

FURTHER READING

(1993) A brighter future for PFBC. *EPRI Journal*, December.

(1993) Advanced combustion turbines set for takeoff. *EPRI Journal*, October/November.

Annual Energy Outlook (1993) US Energy Information Administration, DOE/EIA-0383, Washington, DC.

(1993) Forecast: the electric power industry 1993–2010. *Electrical World*, November.

(1993) Fuel cell demonstration planned. *EPRI Journal*, September.

(1991) Fuel cells for urban power. *EPRI Journal*, September.

Khatib, H. (1993) Electrification for developing countries. *EPRI Journal*, September.

Preston, G.T. (1993) Collaborative Power Technology Development in a Global Business Environment. International Joint Power Generation Conference, Kansas City.

Preston, G.T. (1993) Evolution of Domestic Utility Market Structure into the 21st Century. US DOE Second Annual Clean Coal Technology Conference, Atlanta.

Preston, G.T. (1993) Improved Coal Technologies for the Asia Pacific 21st Century Economic Development. IEA Second International Conference on the Clean and Efficient Use of Coal and Lignite, Hong Kong.

Preston, G.T. (1994) Sustainable Development Through Electricity: US and Global Perspective. Imatron Voima Oy Annual Research Seminar, Helsinki.

WEC Commission (1993) *Energy for Tomorrow's World*. St. Martin's Press, New York.

5

Coal-fired, combined cycle power generation for the 21st century

K.K. Pillai

5.1 INTRODUCTION

Less than 30% of the world population consumes more than 70% of the electricity generated. Around 40% of this electricity generation uses coal. However, only some 10% of the world's coal production is traded on the international market. It seems inevitable that the satisfaction of the aspirations of the major part of the world population requires the use of coal – and primarily domestic coal – for the foreseeable future.

Protection of the environment requires that such coal utilization be with 'clean coal' technologies. In addition the privatisation wave sweeping through the power generation world is a powerful driving force towards lower investment costs (size, modularization and standardization) and lower operating costs (high efficiency, automatic control). The perceived global warming risks – whether real or not – emphasize the importance of thermal efficiency. These are now the driving forces which set the trends.

The environmental driving forces (control of emissions of dust, SO_x, NO_x, CO, carcinogenics, halogens and heavy metals) lead to add-on emissions control for conventional pulverized coal (PC) combustion or the use of the new 'clean' technologies (fluidized bed combustion with inherent emissions control or coal gasification followed by gas scrubbing before combustion). The pressures on investment costs lead to pressurized systems using standardized gas-turbine based combined cycles to be preferred. For the next century these trends point to the use of the new 'clean coal' technologies, using combined cycles for high efficiency. Such plants can be expected to be highly standardized, modularized and automated.

In power generation, the gestation period for new technologies – from laboratory or bench-scale to commercially operating plants – is typically 20–25 years. Pressurized fluidized bed combined-cycle (PFBC) plants are now commercially available. The Värtan PFBC plant in Stockholm is in commercial operation. Two demonstration plants are operating in the United States and Spain. The Wakamatsu PFBC plant in Japan will begin commissioning this year, and a number of other plants have been ordered. The first commercial 350 MW_e plant will begin operation in 1997.

Integrated gasification combined-cycle (IGCC) plants will soon be demonstrated on a commercial power plant scale, beginning with the Buggenum plant in The Netherlands. Advanced PC plants are already operating in several countries. There are plans for installation of PC plants with even more advanced steam data before the end of the 20th century.

While it may well be that fuel cells or magnetohydrodynamics could achieve the technical breakthroughs they need during the next century, commercial power plants using these technologies cannot be considered for the foreseeable future. This chapter concentrates on PFBC and IGCC, which together with conventional PC plants fitted with emissions control are the technologies available for the early part of the 21st century.

World demand for energy, and especially for electricity, will grow substantially during the 21st century. This follows from the increase in world population and the necessary and inevitable economic growth in countries that are currently less developed. Coal – and primarily domestic coal but with an increasing amount of imported coal – will continue to provide for a large portion of the electricity demand. Estimates of the global market for power generation equipment suggest that about 800 GW_e will be needed just during the next decade and that approximately half of that will be coal based.

Pulverized-coal-fired power plants currently dominate the existing power plant spectrum, and have done so since the early part of the 20th century. Continuous improvements have taken place as regards reliability, efficiency, cost and now emissions control. The major efficiency improvements took place between the late 1940s and the middle of the 1960s. The majority of the PC plants, world-wide, have no emission control equipment other than some particulate removal system. Many also have a low thermal efficiency as a result of being old, small, or because they burn low-grade coal. Generally, plants that were built in the early 1950s use about 25–30% more fuel than a modern plant built today. Materials technology limitations in the 1970s led to efficiency reaching a plateau and the emissions control measures were costly – both in investment cost and energy use.

During the 1980s, increased concerns about environmental matters have drastically changed the performance requirements for power plants. This has led to the introduction in a number of countries of add-on desulphurization (de-SO_x) and denitrification (de-NO_x) equipment, with the accompanying increases in power plant cost, size and complexity.

During the same period, major steps have been taken in the development and demonstration of new types of highly efficient, coal-based combined cycle plants, i.e. pressurized fluidized bed combined-cycle (PFBC) and integrated gasification combined-cycle (IGCC) plants. More advanced PC plants have also been designed and recently put into operation.

The recognition that human activities may cause a global climate change has raised general questions regarding the use of fossil fuels as a source of energy. Coal has received special attention since more carbon dioxide, which is one of the 'greenhouse' gases, is formed per unit of energy when coal is burned than when oil or natural gas is used as fuel.

Although global coal usage for electricity production accounts for only a small fraction of anthropogenic global carbon dioxide release to the atmosphere, and there is still no real understanding of the relations between the changes in composition of the atmosphere and the global climate, the mere perception that fossil fuels may cause climate change is likely to influence the choice of power generation technology and also to some extent the choice of fuels. Political decisions in a number of countries have already begun to cause such changes, typically through the use of taxes or fees. For example, in Sweden the cost of importing coal is only about one-third the cost paid by an electricity generator – the remainder being various taxes and environmental fees.

The change will clearly be towards plants with higher thermal efficiencies. It is therefore of considerable interest to compare the most promising coal power generation technologies with respect to efficiency and environmental performance.

This chapter compares efficiency data quoted for modern coal-fired power plants, and seeks to provide a level perspective regarding efficiencies and emissions of sulphur and nitrogen oxides from existing power plants and from new high efficiency 'clean coal power plants'.

5.2 THE DEVELOPMENT OF POWER GENERATION TECHNOLOGIES

Efforts to develop new power generation technologies for use with coal, and to improve on already existing technologies have blossomed through from the early 1970s and are under way in a number of countries. Figure 5.1 illustrates this, in the form of a tree with many branches. The acronyms shown on the different branches of the tree are often found in energy literature. Highly efficient PFBC, IGCC and also advanced PC plants are now being applied on a commercial power plant scale. They will clearly develop further and are likely to form the basis for coal-fired plant well into the next century.

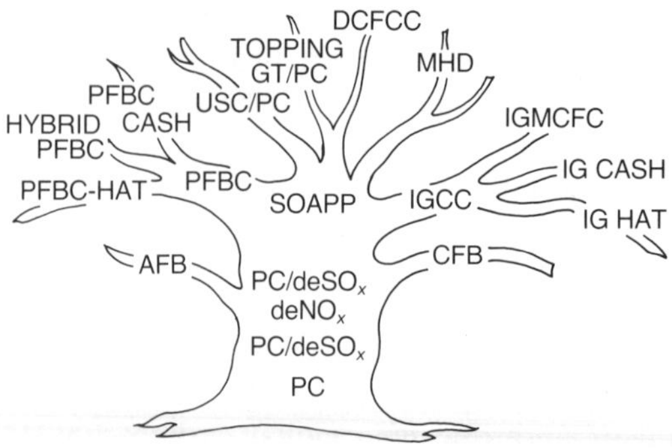

Figure 5.1 The 'development tree' for coal-based power generation.

5.2.1 PFBC plants

PFBC shares the features of fuel flexibility and 'built-in' de-SO_x and de-NO_x functions with the atmospheric fluidized bed combustion plants. But the major advantage with PFBC is the higher thermal efficiency that results from the combination of a gas and a steam turbine cycle. Another is the compactness of a PFBC plant, which reduces the amount of structural steel, and which also allows considerable shop fabrication. This compactness also allows site repowering, i.e. the reuse of old sites, a fact which can be especially important at locations where space is limited or where it is difficult to get permits for new utility sites. The small size of PFBC plants also provides a potential for cost reductions through modularization and reduction of civil works requirements and of construction time.

The properties of the solids residue, which consists of coal ash mixed with partially sulphated limestone, are also favourable and should allow its use for construction purposes. The residue can thus be a resource, which has a positive value and can be sold.

ABB has led the commercial development of PFBC, and is currently offering such plants to utilities and industries around the world in module sizes of 80 and 350 MW_e or multiples thereof. These two module sizes are designated the P200 and the P800, respectively.

A 'first generation' PFBC plant is in commercial operation at the Värtan power plant in Stockholm, Sweden. This is a combined heat and power (CHP) plant which generates 135 MW_e and also 224 MW of heat, which is used in the city's district heating system.

Figure 5.2 shows an aerial view of the Värtan power plant with the PFBC plant at the centre. Besides construction of the new turnkey PFBC plant, the contract also involved the removal of an old boiler plant, major changes to the district heating connections in the plant area, the modification of an

Figure 5.2 Aerial view of the Värtan power plant with the PFBC units (at centre).

underground oil storage rock cavern into a new coal storage, and the construction of coal receiving and preparation buildings in the harbour area.

The Värtan PFBC plant operates during the heating season, which begins in October and lasts to April each year. Through April 1993, the two P200 PFBC units at Värtan have accumulated a total of about 14 500 operating hours, and the plant has now reached a high availability rating. The emissions from this plant are extremely low, or close to what one would normally expect to be possible only from a state-of-the-art natural-gas-fired plant with the best de-NO_x equipment. This is a result of the very strict emission limits that were defined for the plant, which is situated close to the centre of the Swedish capital. This emissions performance is of crucial importance in demonstrating how clean 'clean coal' can be.

The Tidd and Escatron PFBC demonstration plants, which each include one P200 PFBC module, are operating in the United States and Spain, respectively. Still another demonstration plant, based on the P200 module, is now being commissioned in Japan, at the Wakamatsu site of Electric Power Development Company (EPDC). Figure 5.3 shows the arrival of the shop-fabricated combustor vessel, on a barge, to the Wakamatsu site on the island of Kyushu. As in the United States and Spain, the Wakamatsu demonstration project is a preparation for the installation of larger PFBC plants.

Figure 5.3 Photo of the combustor for the Wakamatsu PFBC plant on arrival at the site. The combustor had been shop fabricated and was ready for installation within hours in the boiler house.

Operation in the existing plants has shown that it is, indeed, possible to operate the specially designed GT35P gas turbine in the combustion gas environment, following particulate clean-up only by cyclones. As a result of the successful operating experience at the Värtan plant, a number of PFBC plants will be built during the 1990s. These include a commercial 350 MW_e PFBC plant which will be built at the Karita site of Kyushu Electric Power Co. in Japan, for commercial operation in 1997. This plant, which uses the P800 PFBC module, will have a net thermal efficiency of about 42% (HHV). There are also plans for the construction of P800 size PFBC plants in the United States and in Spain, and engineering studies are under way. A contract has been signed and financing arranged for the supply of a PFBC plant to the Czech utility MST. This will be a CHP plant that produces 60 MW_e, 100 MW heat to the district heating grid and also 10 MW process steam to a neighbouring industry. This contract represents the first new power plant to be financed in one of the former Central Economies in Central Europe.

The PFBC plants mentioned above are all based on ABB technology and involve the use of the P200 and P800 modules, with the specially developed

GT35P and GT140P gas turbines, respectively. It is a measure of the acceptance of PFBC as a viable technology that at least six other equipment suppliers now are actively marketing their designs of PFBC plant to the utility industry. So far, plans to build three PFBC plants with the technology of other equipment suppliers have been announced.

Development is under way to further enhance the efficiency of PFBC plants, through the incorporation of hybrid concepts. Fully coal-fired hybrid PFBC plants include a gasifier or mild gasifier which produces a fuel gas. This gas is used to raise the gas turbine inlet temperature. Other efficiency increases can also be achieved through the use of more advanced steam data.

5.2.2 IGCC plants

IGCC cycles exist in different versions, particularly due to the type of gasifier and the degree of integration between the fuel gas production, the chemical processing part of the plant for 'clean-up' and the power generation equipment. The choice of gasifier depends on fuel quality and, in turn, influences the fuel flexibility of the plant. Together with the degree of integration, it also determines cycle efficiency.

As in the case of PFBC there are now several companies that offer pressurized gasification technology. A major difference is that the companies which have developed gasification technology (generally in the petrochemical world since gasification was originally developed for using coal instead of oil) typically wish to license that technology directly to the utilities that will own the plants. Since utilities are not familiar with that approach it seems probable that in the future there will be licensing arrangements with equipment suppliers instead.

IGCC is prompted as a very 'clean' technology. With wet pressurized removal of hydrogen sulphide and other sulphur compounds that form during gasification, and by using processes well known in the petroleum, refining and chemicals industries, the fuel gas can indeed be made essentially free from sulphur compounds. As a result, sulphur dioxide emissions from IGCC plants can be very low. Similarly, hydrogen cyanide, ammonia and other nitrogen oxide precursors can also be removed through chemical treament before the fuel gas is burnt in the gas turbine combustor. This helps to lower NO_x emissions provided the flame temperature in the combustor is low enough. The disadvantage here is that there is some loss of efficiency related to the cooling down of the fuel gas for clean-up before combustion.

The first utility-scale demonstration IGCC power plant in the world is Demkolec's 253 MW_e. Buggenum plant. It is now under construction in The Netherlands and is scheduled for commissioning in 1994. It uses an oxygen-blown, entrained-flow gasifier of Shell's design, with dry coal feed.

The residual nitrogen from the air separation plant is mixed in with the coal gas before combustion in the gas turbine combustor in order to reduce NO_x emissions. The nitrogen is also used for pneumatic transport of the coal feed into the gasifier. The gas turbine used at Buggenum has a 1050 °C (ISO definition) inlet temperature, and the net thermal efficiency is 43% (LHV).

Several other IGCC projects have been decided on. In Europe there are two other demonstration projects. One is Elcogas's 300 MW_e. Puertollano project in Spain, which is co-funded by a number of utilities from several countries and which has also received European Community financial support. This will have an oxygen-blown Prenflo entrained-flow gasifier, again with dry feed, and a gas turbine with a 1160 °C (ISO definition) inlet temperature. The plant will have a 45% (LHV) net thermal efficiency, and is scheduled to begin operation on coal in 1997.

The other is the 312 MW_e RWE/Rheinbraun Kobra project in Germany, which will use an air-blown high temperature Winkler (HTW) fluidized bed gasifier to produce fuel gas from brown coal. Also in this case, a gas turbine with a 1160 °C (ISO definition) inlet temperature will be used. The char that remains after gasification will be fed to atmospheric fluidized bed combustors to achieve sufficient burn-out and to recover energy. In other words, this is a hybrid type plant with both gasification and combustion employed. The net thermal efficiency is projected to be 46% (LHV), thanks also to a large degree to the use of a fluidized bed brown-coal drier that recovers the heat from condensation of the water contained in the fuel. The Kobra plant will begin commercial operation by the end of 1996.

A number of IGCC projects and two PFBC projects are also planned in the United States, with support from the US Clean Coal Technology Demonstration Program. Most of those IGCC projects are expected to demonstrate hot gas chemical clean-up of the fuel gas, in order to achieve further improvements in plant thermal efficiencies.

5.2.3 PC plants

The major possibility for improving the thermal efficiencies of PC plants from today's state-of-the-art level, which are commonly said to be in the 36–40% range without further qualification of the numbers, is to use more advanced steam data. This requires further increases of steam pressures and temperatures, and ultimately the adoption of ultra supercritical (USC) steam data. The use of double reheat and optimization of the steam cycle also helps push the efficiency figures upwards.

Efficiency levels between 43% and 45% (LHV) have already been reached in PC plants with advanced steam data and low condenser pressure (high vacuum) in Denmark, The Netherlands and Germany. Two 400 MW_e double reheat units with quoted net thermal efficiency of 47% (LHV), at a main steam pressure of 290 bar and with a main steam temperature of 582 °C, will be commissioned in Denmark in 1998.

In Japan, there is a programme to build a number of advanced PC plants before the end of the 20th century with unit sizes up to 1000 MW_e. The first USC/PC plant, the 500 MW_e Tsuruga No. 1 unit, began operation in 1991 and has been reported to have a net thermal efficiency of over 42% on a higher heating value (HHV) basis. A 1000 MW_e USC/PC plant with a main steam pressure of 246 bar and 593 °C steam temperature is scheduled for commissioning in 1997.

Efficiency improvements in PC plants can also be made through the combination with a gas turbine cycle (TOPPING GT/PC). If natural gas is used to fuel the gas turbine, such a plant may reach an efficiency of over 50% (LHV). With coal as the only fuel, a hybrid plant that includes also a pressurized gasification unit, to produce a fuel gas for the gas turbine, is needed. Naturally, it is also possible to replace the PC boiler with an atmospheric fluidized bed boiler in such a hybrid plant. The so-called 'British Coal Topping Cycle' utilizes that combination.

5.3 THE JUNGLE OF EFFICIENCY COMPARISONS

PFBC and IGCC now attract a large interest and are major candidates for use when clean power generation with high thermal efficiency is required. It also shows that recent development of advanced PC plants has produced plants with efficiencies in the same range. Obviously, such advanced PC plants will also find a wide market in the future.

But how should different efficiency data be compared? The common notion in the Unites States is that modern coal-fired power plants have net thermal efficiencies in the range of 36% to perhaps 38%, if equipped with a de-SO_x system. European plants are frequently said to have efficiencies around 40%. And values up to 47% are now mentioned for advanced PC plants in Northern Europe!

One of the major differences which leads to the confusion with quoted efficiencies is the different use of heating value basis for the fuel. In most of Europe it is common to use the LHV basis. But in the United States, Japan and many other countries, efficiency calculations are made based on the HHV of the fuel.

There is an exception, however. Even in those countries where the efficiencies of coal-based power plants are commonly expressed on an HHV basis, the efficiencies of gas turbines and of combined cycle plants fired with natural gas or oil are commonly expressed on the LHV basis. This has interesting results when different types of plants and fuels are considered. For example, the most modern combined cycle plants that burn natural gas today have efficiencies that approach 55% (LHV). And such plants in the future, with new generations of gas turbines, are expected to reach efficiencies near 60% (LHV). This is, of course, impressive! But it is worth while recognizing that in this case a 55% (LHV) efficiency value in reality corresponds to an efficiency of about 49.5% (HHV).

For bituminous coal-based power plants, the difference between LHV and HHV bases alone means that identical plants, with the same site conditions, may be cited as being about 2% more efficient in Europe than in Japan. If the fuel is lignite, and especially if the lignite contains much water, the difference may be twice that value.

One important factor that greatly influences the efficiency of a power plant is the steam cycle condenser pressure. Since this depends on the temperature of the condenser coolant medium and on the heat transfer characteristics for the condenser, it is obvious that climatic effects will be important. A low condenser pressure (or high vacuum) will make it possible to use a steam turbine low pressure stage that can extract significantly more power than is possible at a higher condenser pressure.

The difference can be worth up to 3% if the comparison refers to seawater cooling in Scandinavia on the one hand with a condenser pressure of 0.02 bar (0.6 inches of Hg), and a US summer cooling tower condition with 0.084 bar (2.5 inches Hg) condenser pressure on the other. It is obvious then that these two factors, together, may result in a 5% difference for 'typical', but not defined, coal-based power plants. In many other countries, condenser pressures of 0.04–0.05 bar may be more common, producing still another 'correction factor' when comparisons need to be made.

The third reason is real differences in steam cycle data, and in the degree of plant optimization. For PC plants with reheat, the efficiency will increase about 0.6 percentage points for every 25 °C increase in main steam temperature, and by about 0.8% when the main steam pressure is increased from 180 to 250 bar. A further increase in pressure to 316 bar will raise the cycle efficiency by approximately 0.4%. And the introduction of double reheat may raise the efficiency by another 0.7%. These are not accurate values, but they are given here to allow gross comparisons between different cycles.

Further confusion is caused by definitions of gas turbine inlet temperatures. The 'firing temperature' concept, which is used by one major US gas turbine manufacturer, relates to the average inlet temperature to the first rotor stage. A 'firing temperature' of 1260 °C (2300 °F), which is commonly referred to in the literature, corresponds to a much lower temperature, perhaps about 1170 °C, on the ISO definition scale. This is a calculated temperature value which takes into account all cooling air in the turbine, but which cannot be directly measured anywhere in the turbine.

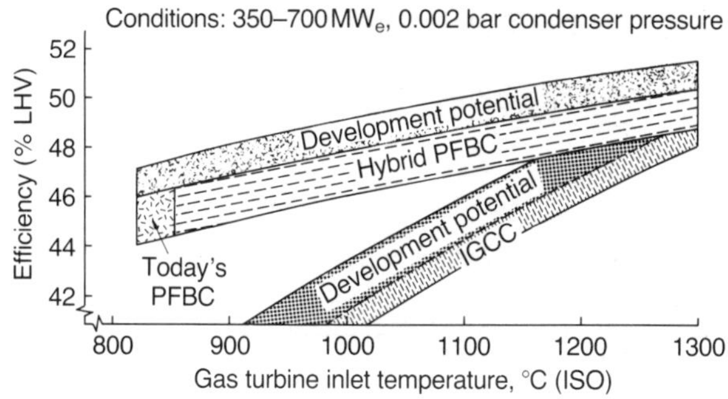

Figure 5.4 Comparison of achievable net thermal efficiencies for PFBC and IGCC power plants on a LHV basis and for a condenser pressure of 0.02 bar.

5.4 COMPARISONS OF EFFICIENCIES OF PFBC, IGCC AND PC POWER PLANTS

Figure 5.4 shows a comparison of efficiency values, calculated on an LHV basis, for utility size PFBC and IGCC plants using typical bituminous coals as fuel. The data represent site conditions which allow a condenser pressure of 0.02 bar, and the efficiency values are shown as a function of gas turbine inlet temperature (ISO definition). The different efficiency 'bands' are valid for well-optimized cycles and show what can be achieved.

The area in Figure 5.4 that is marked 'Today's PFBC' represents gas turbine inlet temperatures of 830–850 °C and reheat steam cycles with main steam pressures ranging from 180 to 250 bar and temperatures between 538 °C and 580 °C. The figure easily allows a comparison with the 43% and 45% net efficiency ranges that will be reached in the Buggenum and Puertollano IGCC plants.

The band marked 'Hybrid PFBC' shows the efficiency improvements that can be reached when the inlet temperature to the gas turbine is increased. For a plant which is all coal based this means including a partial gasifier that provides fuel gas for 'topping' the temperature. There are many different ways the cycle can be arranged to achieve that goal.

The development potential opportunities are also illustrated in Figure 5.4. These relate to improvements in both the gas and the steam cycles. It is interesting to note that the hybrid PFBC and the IGCC cycles tend to reach similar efficiency levels, above 50%, for gas turbine inlet temperatures above 1300 °C (ISO definition). This is expected to happen within the next two decades as a result of further development of materials and turbine blade cooling techniques.

An efficiency comparison for PFBC and PC plants is shown in Figure 5.5, again on an LHV basis and for a low condenser pressure. The major

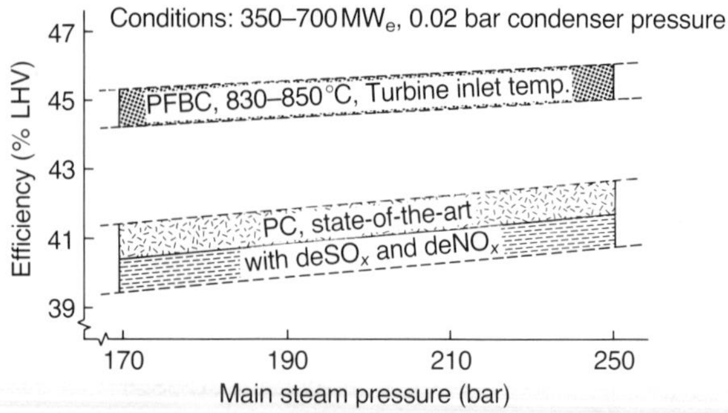

Figure 5.5 Comparison of net thermal efficiencies of PFBS and PC plants on a LHV basis and for a condenser pressure of 0.02 bar.

bands of this diagram represent state-of-the-art PC plants and the PFBC plants that can be built now. The net thermal efficiencies for some of the most efficient PC plants in Europe are also included. It is obvious that these PC plants, and others mentioned above, represent a large improvement on the state-of-the-art technologies. And this has happened during the last few years.

Figure 5.6 shows the achievable efficiencies on a HHV basis and for a condenser pressure of 0.06 bar (1.8 inches Hg). A comparison of the information in Figures 5.5 and 5.6 clearly shows the 4% difference in efficiency numbers that results from a change of heating value basis (about 2%), together with a change in condenser pressure from 0.02 to 0.06 bar (another 2%).

It is also clear that the efficiency value quoted for the Tsuruga No. 1 plant in Japan, although probably for a somewhat lower condenser pressure, approaches what can be achieved with the present PFBC design, which has

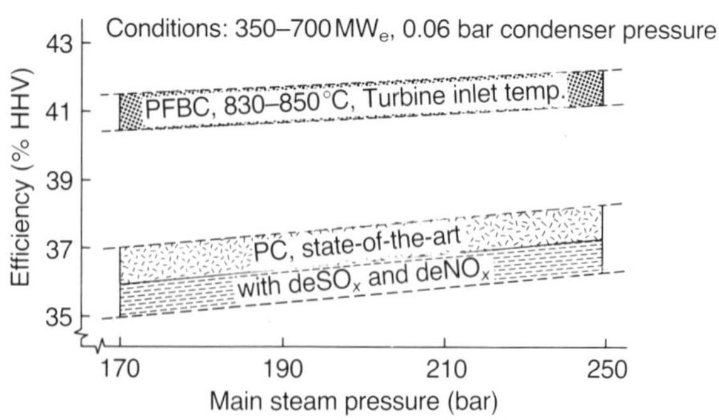

Figure 5.6 Comparison of net thermal efficiencies of PFBC and PC plants on a HHV basis and for a condenser pressure of 0.06 bar.

a potential for development towards higher efficiency values, or with an IGCC plant that uses the most advanced gas turbine that can be made today. Obviously, this situation provides conditions for continued competition between the different technologies. Thus the potential for improvement of the PFBC and IGCC cycles to efficiencies above 50%, as was shown in Figure 5.4, will be important.

5.5 COMPARISONS OF EMISSION DATA

For many years now, it has been the emissions of sulphur dioxide and nitrogen oxides which have attracted major attention. This is because these pollutants are the primary causes of 'acid rain'. Even here, it is difficult to compare information in the open literature because of the variety of units that are prevalent.

Figures 5.7 and 5.8 show diagrams that allow a conversion between different units. The different bases for emissions relate to either the energy

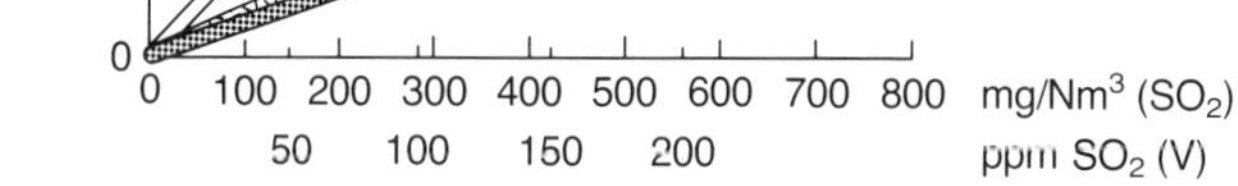

Figure 5.7 The relations between emission values for sulphur dioxide (or sulphur), expressed in different units.

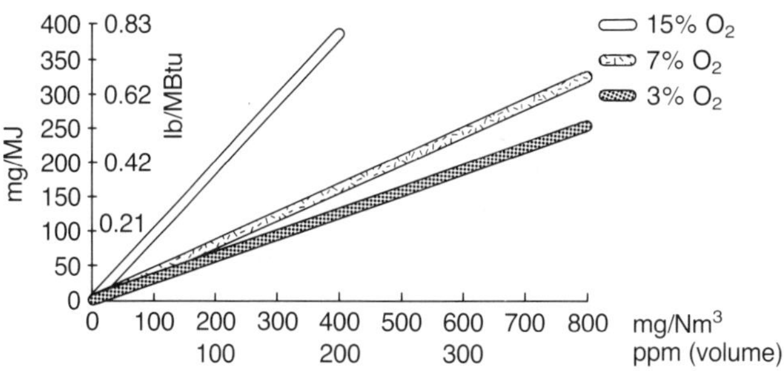

Figure 5.8 The relations between emission values for nitrogen oxides, expressed as NO_2, in different units.

content of the fuel used (vertical scales), or else to the concentration of pollutants in the flue gas stream (horizontal scales).

But there are some complications. Some authorities regulate for sulphur emissions while others quote values for sulphur dioxide. Heating values used in the United States are normally on an HHV basis while in Europe LHV is used. It is also important to recognize that the emission values on a flue gas volume basis normally relate to dry gas. And NO_x is interpreted as meaning NO_2, even though most of the nitrogen oxides considered are released as NO molecules. Nitrous oxide, N_2O or 'laughing gas', however, is not included at all. This is because it has not been a subject of major discussion until recently, when 'greenhouse gas' concerns and protection of the ozone layer have become more of an issue. While this is logical from an environmental impact viewpoint, it is difficult to justify why di-nitrous oxide is not an oxide of nitrogen! It is also important to know the level of oxygen in the flue gas stream when conversions are to be made from 'fuel energy specific' values to 'flue gas volume' specific values, or vice versa. The level of oxygen directly relates to the degree of dilution of the combustion products with air. A 15% oxygen level is typical for modern gas turbines, and also in IGCC plants that use high temperature gas turbines. Those require dilution with air in order to allow expansion through the gas turbine. This cooling occurs both in the gas turbine combustor and inside the actual gas turbine. By comparison, the oxygen levels in flue gases from current PC and PFBC plants are typically kept low in order to minimize 'stack losses'.

5.6 THE FUTURE POWER PLANT SPECTRUM

The spectrum of coal-based power plants in operation throughout the world during the next few decades will undergo a change. For economic reasons most existing power plants will generally continue to be used till the end of their useful 'life'. On average this is likely to be 40 to 50 years. Therefore, most existing capacity will remain in use for another 20 to 30 years, except in cases where changes are forced to occur through legislative or other actions that change the viability and economic attractiveness of such plants.

Second, a number of existing PC plants will clearly be modified to prolong their life and to improve performance. Depending on what plant components need replacement, such rehabilitation may or may not involve repowering. Retrofitting with de-SO_x, and de-NO_x equipment will occur in cases where the existing power plants are otherwise in good shape.

As regards construction of new capacity and repowering of older plants that require major refurbishment, it is obvious that political and market forces will to an increasing degree require such units to have a high thermal efficiency and to have low emissions of environmentally harmful substances. In other words, they must be 'high efficiency, clean coal plants'.

And clearly, PFBC and IGCC plants have an opportunity to be widely implemented.

5.7 CONCLUSION

It is interesting and enlightening to reflect on the relations between efficiencies and emissions from the existing spectrum of power plants, and from the new highly efficient and clean PFBC, IGCC and PC plants. Figure 5.9 shows an estimate of how these improvements may develop with time. It is notable that the world average efficiency will increase only slowly, owing to the sluggishness with which the existing power plants will be upgraded or replaced. This must be seen as a large problem in a situation where it is important to supply increasing amounts of electricity, while limiting the

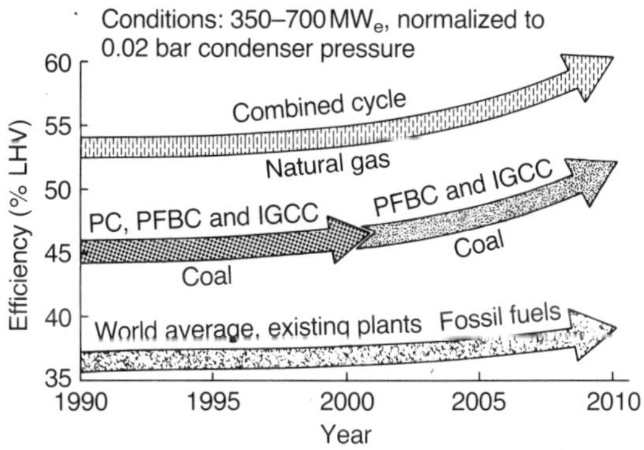

Figure 5.9 A scenario for the efficiency improvements in different types of power plants, and of the world average efficiency for all power plants.

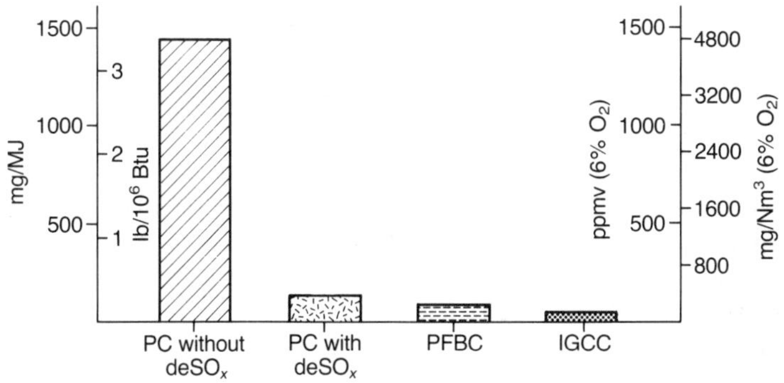

Figure 5.10 Typical sulphur dioxide emissions from coal-based power plants. The data for PC without de-SO_x represent an estimated world-average.

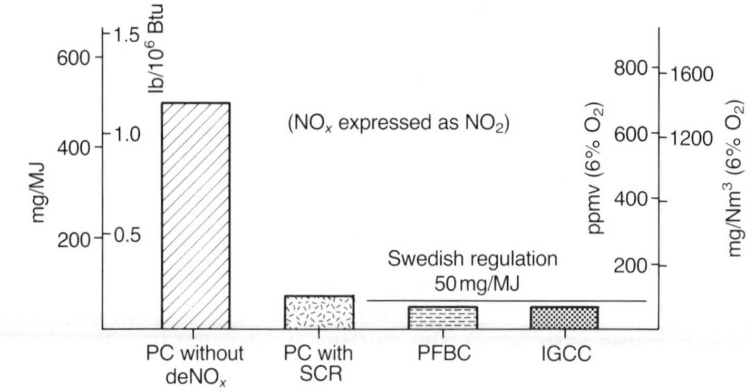

Figure 5.11 Typical nitrogen oxide emissions from coal-based power plants. The data for PC without de-NO$_x$ represent an estimated world-average.

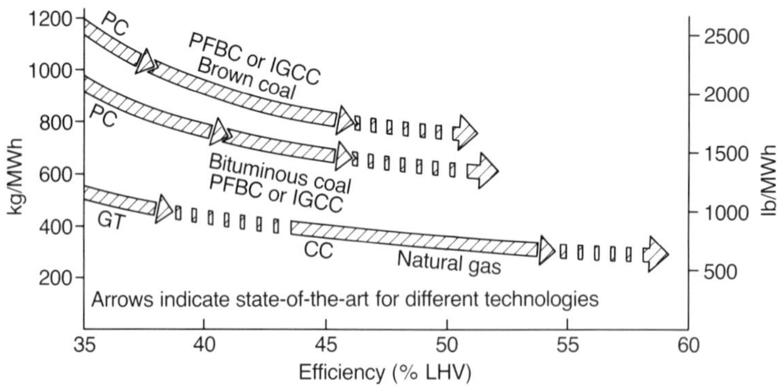

Figure 5.12 Carbon dioxide emissions from power plants with different thermal efficiencies and fuels.

use of energy resources. Environmental impact minimization is clearly served best by the rapid introduction of the latest available technology.

When it comes to emissions of sulphur dioxide and of NO$_x$, it is worth while to compare world-average data with what can typically be achieved with the new technologies. Figures 5.10 and 5.11 provide such comparisons. Again, it must be recognized that most existing power plants in the world have no de-SO$_x$ or de-NO$_x$ capabilities. It is obvious from these figures that emissions from those power plants deserve much attention, whereas the emissions from the 'high-efficiency, clean coal technologies' already approach those that can be achieved from natural-gas-fired plants. Figure 5.12, finally, provides a picture of how the carbon dioxide emissions are affected by changes in the thermal efficiency of power plants burning brown coal, bituminous coal and natural gas.

6

Nuclear power – energy for the 21st century

R. Hawley

6.1 INTRODUCTION

Over the next century the issue which will most concern the world's politicians, industrialists and environmentalists alike will be extending and sustaining the benefits of energy use without increasing the risk of irreversible damage to the environment. An inevitable growth in world-wide energy demand cannot be sustained on fossil fuels alone for they are both finite and a major source of environmental damage.

Nuclear power is one of the solutions to the conflict between energy availability and environmental protection. It is tried and tested, supplies 17% of the world's electricity and has reserves with double the energy potential of fossil fuels. If the world is to realize the huge potential of nuclear energy a number of challenging issues have to be resolved. Are the nuclear and engineering industries capable of building and operating the increasing number of plants that will be needed? Can such an expansion be afforded and the benefits transferred to the developing world? Can the safety of a major programme be assured – in operation, in the management and disposal of the wastes that are produced and in controlling the ever present risk of nuclear weapons proliferation? Above all can the industry achieve the level of public and political trust that will be required to sustain the technology through a period of energy plenty and the dominance of markets that are short term in outlook and do not reward environmental and strategic benefits?

The evidence is that the nuclear industry is capable of delivering all that is required of it, but this will not be enough. Only governments can create the environment which will ensure the required contribution from nuclear power, conservation, renewable technologies and the many other developments that are essential to world-wide sustainable economic development.

The world, its people and their governments today face very serious challenges in respect of energy use and its effects on the environment. These challenges lead to choices that will have far reaching consequences for the stability and even long-term viability of life on our planet.

Energy is vital to the quality of life – a minimum level of energy supply of the equivalent of about 1.5 tonnes of oil per year per capita is necessary to achieve reasonable levels of infant mortality, life expectancy and literacy. Today at least 80% of the world's population live in regions where the average energy supply falls well below this level – and these regions have the highest rate of population growth.

A growing population automatically imposes even greater demands for energy. This applies especially in those areas that are currently under-developed and desperately seek the minimum standards that the developed world now takes for granted.

The fossil fuels that have created the opportunity to improve the situation also, by their very nature, present serious problems.

1. They are finite, challenging the sustainability of energy supply.
2. Their use is polluting, threatening the local and global environment upon which life itself depends.

Other forms of energy, like most renewables, are either too costly, inconvenient and diffuse to be credible substitutes for fossil fuels or, like nuclear power, are identified with their own specific problems and do not have, as yet, willing public approval. The obvious alternative is to use less.

1. Energy efficiency improvements have been sought and realized since energy became the traded commodity underlying all economic development. Since these improvements are market driven, the resultant impact on economic growth has outstripped the economies obtained from efficiency improvement and the demand for energy has continued apace.
2. Conservation is a higher level objective than simple energy efficiency. It is a deliberate attempt to reduce absolute energy consumption. Conservation can be achieved in one of two ways. One approach is self-restraint in energy use. Another is to achieve a substantial reduction in energy intensity that is not driven by market forces and so is not cancelled out by consequential increases in demand. The latter approach has usually required regulation to achieve any degree of success.

The World Energy Council has addressed these issues in its wide ranging analysis of world and regional energy needs over the next 30–100 years (World Energy Council, 1993). The conclusions of this study can be summarized as follows.

1. In all future scenarios that provide for acceptable levels of economic growth there is a substantial growth in the total demand for energy. This

remains true even in an 'ecologically driven' world of sustained rapid reduction in energy intensity.

2. While there is no suggestion of an absolute shortage of energy in the next few decades, the pattern of global demand growth and resource depletion will mean that the geographical distribution of energy resources will not match the distribution of energy demand.
3. Because of expected population growth alone - even with a modest increase in per capita energy consumption to no more than a subsistence level - there will be global energy demand growth of between 30% and 95% by 2020. This will occur almost exclusively in the newly developing countries, whose indigenous supplies are dominated by fossil fuels.
4. It will be virtually impossible to limit man-made CO_2 emissions to 1990 levels by 2020, even assuming the maximum contribution from conservation, energy efficiency, renewables and nuclear power.

Beyond the 2020 horizon, even if environmental concerns have not already driven down the use of fossil fuels, growing scarcity will increase dependence on alternative sources of energy. Table 6.1 shows the scale of growth required in the 21st century. Actions to encourage every aspect of reduced fossil fuel dependence to meet the needs of 2020 have to be taken immediately. Failure to act at once will not only jeopardize economic growth and the global environment but will make meeting the greater challenges of the rest of the century impossible.

6.2 THE CHALLENGES FOR NUCLEAR POWER

The WEC report identifies that electricity demand will continue to grow at a faster rate than energy demand. One driver is that electricity is the key to improved quality of life. It is also the principal route to the exploitation of nuclear, hydro and 'new' renewable resources.

By assuming that nuclear power is predominantly used as a source of electricity, we can build up a picture of the task facing the world's nuclear suppliers and operators. Table 6.2 translates nuclear energy contributions world-wide into the nuclear capacity that has to be in place over the 21st century.

To rise to this challenge a number of issues have to be resolved.

1. Is the industry capable of constructing and operating the number of plants required to meet the target for 2020 and then the even greater expansion implied for the period beyond?
2. Can these programmes be afforded and would they represent the best use of the available capital resource in comparison with alternatives?
3. Can the industry ensure the safety of the large number of plants spread across the world and manage the environmental and waste problems they

Table 6.1 WEC suggested future global energy demand (Gtoe/annum)

WEC case		*'High growth', A*			*Reference, B*			*'Ecologically driven', C*		
Year	*1990*	*2020*	*2050*	*2100*	*2020*	*2050*	*2100*	*2020*	*2050*	*2100*
Fossil fuels	6.80	13.1	15.7	16.8	9.8	13.1	10.9	7.6	8.7	3.0
Nuclear	0.44	1.0	3.8	12.2	0.8	3.5	9.2	0.7	1.2	2.2
'New' renewables	0.18	0.8	4.1	10.1	0.5	3.2	8.6	1.4	3.0	10.0
Traditional renewables and hydro	1.38	2.3	3.4	2.9	2.3	3.2	4.3	1.6	2.1	4.8
Total primary energy demand	8.8	17.2	27	42	13.4	23	33	11.3	15	20

Table 6.2 Number of reactors required to meet the WEC's demands

	2020	*2050*	*2100*
Global energy demand (Gtoe)	11–17	15–27	20–42
Nuclear energy demand (Gtoe)	0.7–1.0	1.2–3.8	2.2–12.2
Nuclear capacity (GW_e)	431–610	746–2352	1369–7581
No. of reactors in operation	308–436	533–1680	978–5415

will create? In particular, can this global expansion be achieved without leading to an unacceptable risk of nuclear proliferation?

4. Considering that the growth in energy demand will be concentrated in developing countries, how can the expansion in nuclear capacity help meet these countries' legitimate aspirations?
5. In addition, there are many barriers to overcome. Some arise from continued adverse public perceptions, which so strongly influence the political acceptance of nuclear power. Others result from the developments taking place in the energy markets by which short-term commercial considerations increasingly dominate investment decisions and strategic and environmental benefits are largely ignored. Can these issues be resolved in time?

6.3 THE INDUSTRIAL CAPABILITY OF THE WORLD NUCLEAR INDUSTRY

6.3.1 The present position

At present nuclear power supplies 5% of the world's primary energy by providing around 17% of the world's electricity demand. At the end of 1992, 424 nuclear reactors were in operation in a total of 29 countries generating more than 2000 TWh of electricity – greater than the total amount of electricity produced from all sources in 1958 when nuclear power was in its infancy.

In seven countries nuclear generation meets more than half of the electricity demand. Overall, nuclear power is the biggest single source of electricity in Europe, providing one third of its electricity needs.

The place where interest in nuclear power is growing is East Asia. Japan has 45 nuclear plants in operation and plans to expand its nuclear capacity with the addition of 20 new plants before 2010. Taiwan, with its demand for electrical power rising by over 6% per annum, has plans to increase its capacity, as do South Korea and China.

6.3.2 Implied demand and construction requirements

What is the ability of the world's nuclear industry to construct the new plant implied by the WEC?

The WEC 'high growth' case has been taken as the upper bound and the 'ecologically driven' case as the lower bound of what is required. In addition, it is assumed that all new plants will be large units, typical of the most recent advanced PWR and BWR, and that they will achieve average load factors of 80% or more.

By the year 2020 there would need to be 310 to 436 reactors in operation. Taking account of the need to replace some existing plant, this implies that between 220 and 400 new large reactors will have to have been brought into service by 2020, predominantly in Europe, America, the Pacific Rim and the Commonwealth of Independent States (CIS) (Table 6.3). Thus, in addition to the 72 plants currently under construction, a minimum of 148 reactors will have to commence construction within the next 20 years. At an average rate of commissioning between 8 and 16 reactors per year, this is well within the capability of existing vendors.

Beyond 2020, the implied rate of construction increases. Between 533 and 1680 large reactors would need to be in service by 2050 and between 980 and 5415 reactors by 2100. The highest rate of commissioning required in the ecologically driven scenario is 25 units per year in 2100. This too is less than has already been achieved historically. The much greater rate of

Table 6.3 Increase in nuclear capacity by 2020
(a) 'Ecologically driven' case

Region	*Increase in nuclear (toe)*	*Increase in nuclear (TWh)*	*New reactors (growth + replacement)*
North America	15	67.2	7 + 46
Latin America	22	98.56	10
Western Europe	55	246.4	25 + 40
Central and Eastern			
Europe	14	62.7	6 + 5
CIS	18	80.6	8 + 21
Middle East/N Africa	0	0	0
Sub-Saharan Africa	4	17.92	2
Pacific	100	448	45
South Asia	24	107.52	11
World	275	1232	226

(b) 'High growth' Case A

Region	*Increase in nuclear (toe)*	*Increase in nuclear (TWh)*	*New reactors (growth + replacement)*
North America	36	161.28	16 +46
Latin America	46	206.08	20
Western Europe	80	358.4	36 + 40
Central and Eastern			
Europe	25	112	11 + 5
CIS	46	206.08	20 + 21
Middle East/N Africa	0	0	0
Sub-Saharan Africa	11	49.28	50
Pacific	214	958.72	96
South Asia	83	371.84	37
World	541	2423.68	398

commissioning required by the 'reference' case reaches 130 reactors per year by 2100. This exceeds previous experience but since Framatome, in France, has previously achieved a rate of five reactors per year continuously over a 10-year period, it cannot be out of the question for this to be repeated by 20 or more vendors world-wide if the opportunity presents itself and the programme is phased to meet production capacities.

'Opportunity' however, may be the key. Although there is evidence of demand-led growth of nuclear generation in the Pacific Rim this is not the case elsewhere. Indeed, in Europe and America it will be extremely difficult even to maintain the present nuclear capacity. During the first quarter of the 21st century some 80 GWe will have to be replaced in Europe alone. At present, only France is committed to building further nuclear plant. In the United Kingdom the ability to maintain existing capacity is totally dependent on the outcome of the forthcoming Nuclear Review. If there is no positive commitment from this Review matched in other parts of Europe and America, the required plant will not be built and the industrial capability outside of France and the Pacific Rim will almost certainly wither.

In the author's view, 130 reactors per year world-wide, and more, is physically achievable. Further, as an optimist, the author believes that the West's nuclear capability will not be allowed to die, so on to the next question.

6.4 IS A MAJOR NUCLEAR PROGRAMME AFFORDABLE?

There is a popular perception that nuclear energy is particularly capital intensive and a programme of any significant size could not be afforded.

The reality is that there is nothing unique about the capital requirements for nuclear power except in the scale of individual power station projects, the distribution of the capital burden and the payback period involved.

In making the comparison of capital requirements for a whole spectrum of generating systems, I have taken an unfamiliar starting assumption - all primary energy is free! The wind is free, the sun is free but so are coal, oil, gas and uranium. In all cases costs only arise when you exploit the free resource. There can be no coal-fired power stations without mines, no CCGTs without gas production, no wind energy without appropriately sited wind turbines.

Table 6.4 gives a very much simplified overview of the capital required to exploit the fuel resource and to convert it into electricity. In every case the quantities refer to the same annual energy output, the 19 TWh that would be produced by a large modern twin PWR.

Coal, oil and gas have very similar capital requirements, in the range £3.2 billion to £3.8 billion. When the expected lifetime is taken into account, the balance is tilted in favour of nuclear because of the long life experienced and expected from modern PWR and BWR plants. Gas, surprisingly, is significantly more capital intensive than either nuclear or coal.

So why is it that in many countries, including the United Kingdom, gas-fired CCGT is a more attractive investment than nuclear? Ignoring, for now, the public acceptance and political dimensions and their effect on investor perceptions, it is the way in which the total investment is distributed and the distribution of commercial risk that provides the clue. The notional gas field would not in fact be dedicated to one set of CCGT plant over its 25-year operating life. Achieving the highest possible rate of return requires the fastest depletion of the field and, typically, its output would all be traded to a variety of customers over around 10 years. However, the CCGT owner is not exposed to a significant risk either. While there are ample reserves, he can expect to be able to contract for further supplies from new gas fields developed later and recover the new price in the electricity market. In the unlikely event that the electricity price is insufficient, he can still trade his gas to other customers. In contrast, for nuclear power, although there is a reasonable choice of competitive fuel suppliers they, and the nuclear power station operator, are totally dependent on income from one source - electricity sales. Without some form of market guarantee this leaves the whole of the nuclear industry exposed to a significant level of risk.

So market circumstances make gas-fired CCGTs preferable to coal and nuclear power and, while gas is readily available, the coal and nuclear industries are at serious risk of decline. Yet it is clear in the author's view that in terms of their total demand for capital, there is no major difference between gas, coal and nuclear technologies but investment in generation from renewables is 2 to 15 times more capital intensive for the equivalent

Table 6.4 Capital requirements for the annual energy equivalent of a twin PWR power station (19 TWh/year)

	Fossil fuel		*Nuclear*	*'New renewables'*		
	Coal	*CCGT*	*PVR*	*Wind*	*'Severn' tidal*	*Solar*
Lifetime load factor (%)	80	80	83	29	24	15[a]
Annual energy (Twh)	19	19	19	19	19	19
Required capacity (GW)	2700	2700	2614	7480	9037	14460
Number of 'units'	2	8	2	15–25 000	1	7
Life (years)	30	25	40	20	>50	30
Lifetime fuel required	214 Mt	90 bcm	10–16 000 tU[b]	–	–	–
Plant capital (£ billion)	£2.2	£1.4	£3.5[c]	£7.5	£10.6	£47
Fuel and indirect capital (£ million)	£1000.0[d]	£2000.0[e]	£50–300[f]	–[g]	–[g]	–[g]
Total capital (£ billion)	£3.2	£3.4	£3.5–3.8	>£7.5	>£10.6	>£47
Capital/TWh over life (£ million)	£5.6	£7.16	£4.6–5.0	>£19.7	<£11.1	>£82.5

[a] Load factor based on UK – better than this in lower latitudes.
[b] As natural uranium, corresponds to 2000 tonnes of enriched fuel.
[c] Based on Sizewell C. French experience with series construction would reduce this.
[d] Assumes modern deep mine. Open cast mining would result in a much lower value.
[e] Assumes the undiscounted cost of a series of medium sized gas fields.
[f] The higher figure assumes reprocessing and MOX recycle.
[g] Excludes the indirect capital cost to provide firm capacity.

energy output and cannot possibly succeed in the market without massive financial support. If the availability of capital is an issue, as it surely must be in the developing world, it is hard to believe that anything other than the use of readily accessible fossil fuels or, with help, nuclear power will be affordable for electricity generation. Reliance on renewables may be affordable in small quantities but the scale proposed by the WEC would represent a very high price to pay for the avoidance of perceived risks of a greater use of nuclear power.

As France has already demonstrated, with serious and sustained political commitment, nuclear power is not only affordable, it brings genuine competitive advantage to energy customers.

6.5 URANIUM AND PLUTONIUM RESOURCE

Although not of use for any other purpose, uranium is also a finite resource. It is a quite common element but, as with all minerals, the resource base only takes account of reserves that can be exploited within a price that the market has set. On this basis, the known reserves are some 4 megatonnes (at US$130/kg). This corresponds to around 60 years of supply at the current rate of consumption and with existing, essentially once through, thermal reactor fuel cycles. However, the International Atomic Energy Agency estimates that there is up to four times this amount yet to be discovered. This is just sufficient to sustain the nuclear programme implied by the WEC's 'ecologically driven' case to the end of the 21st century. In all other cases, the proven and speculative resource is sufficient only to sustain the implied reactor programme up to about 2030–2050.

However, there is every reason to expect that, just as with oil, gas and minerals in general, clearly signalled growth in demand will encourage further exploration and discovery. However, it would be unwise to rely solely on successful exploration to meet the growth in demand, nor is that necessary. There is already a lot of experience of recycling uranium and plutonium recovered from spent fuel as MOX (mixed oxide) fuel in light-water reactors, extracting around 50% more energy from the uranium resource alone.

Beyond that, of course, is the fast breeder reactor which has the capability of extending the realizable energy potential of the original uranium by at least 50 times. With fast reactors, the known and 'reasonably assured' reserves of uranium exceed the total reserves of all fossil fuels and could last for over 1000 years.

6.5.1 Proliferation

One major barrier to either route to extending the uranium resource base is the current international government and public concern that reprocessing and the separation of reactor plutonium will increase the risk of nuclear weapons proliferation. The author takes the view that nuclear power is not the problem, but the key to the solution.

The end of the cold war has yielded a vast quantity of highly enriched uranium and plutonium, from surplus nuclear weapons. The materials exist and the knowledge to generate more cannot be destroyed. Both are a major cause for concern and are a clear indication that proliferation will always need to be taken seriously whether or not there is nuclear power generation into the future. Proliferation can only be prevented by a strict international safeguards regime, established by governments, implemented by the International Atomic Energy Agency and enforced by the United Nations.

Today there is a decision to be taken over the future of these materials – store them securely for millennia or burn them? The second must be the right choice.

All commercial reactors burn plutonium. Even in the majority that do not use MOX fuel, around one-third of all the heat generated is the result of the fission of some of the plutonium produced during operation. Current stocks of highly enriched uranium and plutonium can be readily diluted with natural or recycled uranium to achieve the low enrichments suitable for power reactor use. The estimated 1500 tonnes of highly enriched uranium would dilute to around 55 000 tonnes of standard enrichment PWR fuel (1880 reactor years of feed, 8 to 9 years of supply for current PWRs). With the standard MOX cycle (one third core), the total Russian and US inventories of surplus plutonium, estimated to be 200 tonnes, could supply about three years of feed fuel for the world's PWRs. This puts the problem into perspective.

There is nothing new in the use of plutonium for power generation in thermal reactors and the nuclear industry can easily absorb all existing and new plutonium from weapons and civil reprocessing sources and render it unavailable for weapons use. Its use in this way avoids the proliferation risk so long as the relatively small stocks of separated materials in process between reprocessing and recycling are properly safeguarded. Since there is no reason why both activities should not take place on the same site this should also provide the most straightforward safeguards route into the future.

6.6 SAFETY AND ENVIRONMENTAL ISSUES FOR NUCLEAR POWER

It cannot be denied that nuclear power carries unique environmental disadvantages which, for many, are a sufficient justification for opposing its use at all. The process of nuclear fission produces a series of solid and gaseous fission products that are highly radioactive and have half-lives ranging from hours to many thousands of years. Exposure to these materials can represent a direct risk to health and, in the extreme, to life. Avoidance of the hazard requires a series of very effective phases of safety management.

First, there is the need to contain these materials during reactor operation, through the control of routine discharges, the prevention of accidents and the containment of any accidental releases that could occur. Second, there is the need to manage the subsequent storage of spent fuel until it is reprocessed or disposed of. Third, there is the need to contain these materials during and after any reprocessing cycle, to separate the reusable materials from the true wastes. Finally, all the wastes must be disposed of in such a way as to effectively eliminate the risk of their return to the ecosystem.

Can this be achieved with around 5500 reactors in operation world-wide?

6.6.1 Operational safety

The record of safe operation achieved by the nuclear industry, even including the disastrous accident at Chernobyl, is unrivalled by any other section of the energy industry. The reason is that the potential for harm arising from radiation exposure was well known long before the first demonstration of a controlled fission reaction in 1942. Then and at all times since, the primary aim of designers and utilities has been to protect operators and the public from radiation exposure and possible harm.

Of course, the industry has not always succeeded but the priority that is afforded to safety has ensured that the industry world-wide learns from each and every event wherever it occurs. The fire in an open cycle air cooled military pile at Windscale in 1957 taught us the vital importance of effective containment and control of discharge routes. When the reactor core was severely damaged in the Three Mile Island accident in 1979, the event was a commercial disaster but fully contained. The consequences were a greater focus on the role of the operator and his information and training needs. The establishment of the Institute of Nuclear Power Operations (INPO) facilitates the exchange of operating experience and the promotion of best practice. The former Soviet Union was not initially involved but Chernobyl reminded us that nuclear power is a global issue. Through the formation of the World Association of Nuclear Operators (WANO), international co-operation and the promotion of the highest standards of operational safety and reliability are now truly global.

The current level of international co-operation between utilities, regulators and vendors holds out the prospect for the international harmonization of safety objectives and assessment criteria. This would bring benefits to everyone. One benefit would be uniformly high standards of safety in design and operation that would contribute to real gains in overall safety and the restoration of confidence in the nuclear industry.

Another would be the opening of the market for nuclear plant with the minimum requirement for expensive design change simply to satisfy some special national regulatory requirement.

The ultimate goal would be a form of design certification similar to that established in the aircraft industry. In such a regime, a nuclear island that had been certificated in one participating country would be acceptable for licensing by suitably qualified and regulated operators everywhere. This would enable nuclear power to achieve its optimum cost competitiveness and transferability without compromising safety. The necessary precondition is an international agreement on high level safety requirements which, increasingly, are probabilistic rather than deterministic.

The 'tolerability of risk' framework developed in the United Kingdom is one example of how probabilistic criteria can be developed. Risks are divided into three bands: unacceptable, tolerable and broadly acceptable. Within this framework plant would not be licensed if its risks lay in the unacceptable region. In the tolerable region plant is, in principle, licensable, but the ALARP requirement of UK law requires the risk to be pushed down to as low a level as is reasonably practicable. Only in the broadly acceptable region, which demands far higher levels of protection from nuclear risks than any other, would the regulator generally not pursue the operator to reduce risks further.

Although this framework is becoming more widely recognized, individual numerical data have yet to be agreed. But a consensus is growing that the frequency of substantial core damage should be less than 10^{-5} per reactor per year and the frequency of a significant off-site release, which would require limited off-site countermeasures, at least an order of magnitude lower than this, at 10^{-6}. This implies that if there is a large number of reactors there will inevitably be a small number of incidents that result in serious plant damage. With 5500 reactors in operation serious plant damage may occur up to once every 20 years world-wide with a 'significant' release up to once every 200 years.

Of course it is difficult for people to understand and believe such claims, or even to accept that one significant release per 200 years is really tolerable. For this reason, the industry recognizes that any breach of safety represents a serious commercial threat to the industry as a whole. All operators are very strongly motivated to avoid any form of safety risk and, through ever stronger international co-operation in all aspects of plant design, operation and maintenance. Our aim is to achieve even higher standards in practice, leading to more efficient operations, even fewer incidents giving rise to economic damage, and to ensure that any future 'serious' event will be so well contained that a 'significant' release is more akin to the experience of Three Mile Island than Chernobyl.

This aim closely follows the general experience in all branches of engineering (Figure 6.1) except that nuclear power starts from a level of

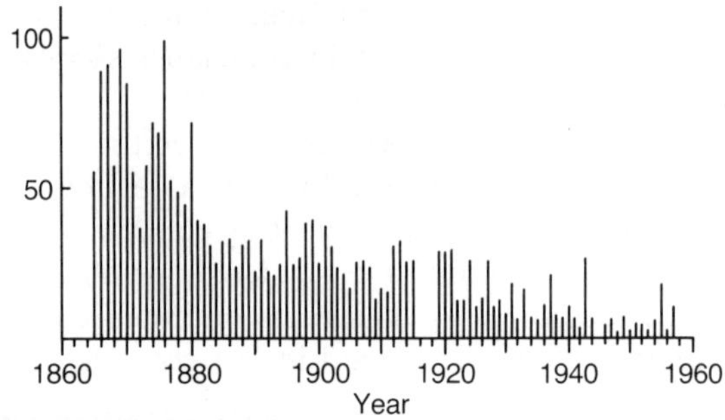

(a)

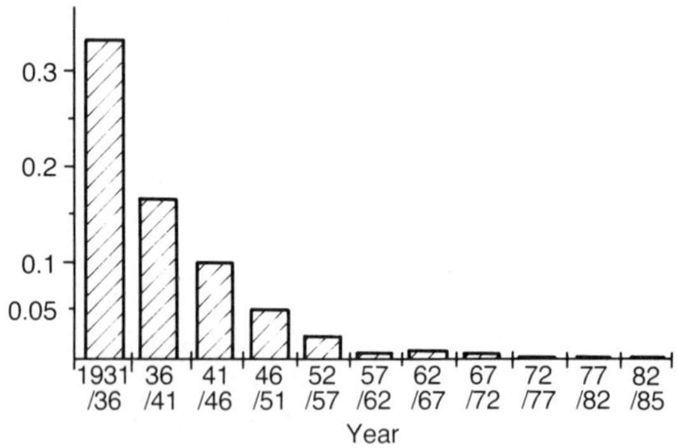

Figure 6.1 (a) Persons killed by boiler explosions in the United Kingdom. (b) Passengers killed per million flying miles (United Kingdom).

safety achievement and public protection that most others have yet to match.

6.6.2 Decommissioning

The only thing that distinguishes nuclear plant decommissioning from the dismantling of any other plant is the presence of radioactive contamination. This contamination leads to a need for special precautions in just the same way as when dismantling structures containing asbestos or toxic chemicals. But there is a difference that is not widely appreciated.

There is a growing consensus that, because the level of radioactivity reduces naturally quite quickly, decommissioning of nuclear facilities becomes simpler with time and the level of risk to those undertaking the work is reduced. This is quite the reverse of the normal experience with chemical and other hazards where the material does not become safer over time and its condition and containment become less predictable.

However, this logical approach does not as yet satisfy many of the public who, understandably, cannot ignore the suggestions from some quarters that the policy is driven by a mixture of inability to actually achieve decommissioning or to immediately meet the cost. The real problem reduces to a lack of trust in the industry's judgement and motives. The industry world-wide is only now really coming to terms with the fact that trust is not won by scientific argument and technical debate but by convincing actions that build a positive track record. This was made clear by the Royal Commission on Environmental Pollution as long ago as 1976 (Flowers, 1976).

Therefore, the complete dismantling of the reactor at Shippingport in the United States is promoted strongly as evidence of what can be done. In the United Kingdom joint action to dismantle the Windscale prototype AGR (WAGR) will be open to public scrutiny. Nuclear Electric is currently dismantling two nuclear plants, Berkeley and Trawsfynydd, and Scottish Nuclear is decommissioning Hunterston A. These will be valuable demonstrations and will prove to the public what can be done.

6.6.3 Waste management and disposal

The management of radioactive waste is another area where the industry is confident of its ability to provide all that is required. Even so, public concern is focused on the final disposal of long-lived high and intermediate level wastes.

Every industrialized nation has to deal with some low and intermediate level wastes that arise from medical and industrial sources. However, only those countries with nuclear reactors have to deal with high level waste, whether in the form of spent fuel or the treated waste stream from reprocessing. Even so, in volume terms, the quantities of nuclear waste that are involved are minute compared with the general level of toxic and other waste production by industry. In the European Union, which has the highest proportion of nuclear power of any region in the world, around 80 000 m^3 of radioactive waste are produced each year, of which only 150 m^3 are highly radioactive. This compares with 10^7 m^3 of toxic industrial waste and 10^9 m^3 of industrial waste of all types.

Each nation has independently reached the same conclusions on how to dispose of its wastes. Shallow burial is all that is required for low level and short-lived intermediate level wastes. Suitable disposal sites are in service in at least nine countries, including seven in Europe. Because of this experience low level waste disposal is not now the subject of public concern.

Long-lived intermediate level waste and high level waste will eventually have to be disposed of in deep underground repositories. There is one deep repository under construction at Gorleben in Germany and at least six others at the planning stage. In several cases, including the UK NIREX repository, work is being preceded by large-scale investigations in so-called 'rock characteristic laboratories'.

High level waste has the special characteristic that in the first few years following its production the rate of radioactive decay makes it strongly heat generative. This waste is best managed by a period of interim storage where it can be cooled in a controlled way for 50 or more years, after which it can be placed in a deep repository.

The small volumes of waste that have to be disposed of make it possible to control the containment environment economically to provide a series of barriers. The ways in which active material can migrate through these barriers and away from the disposal site can be modelled in discrete phases and can take advantage of directly relevant experience from nature going back almost 2000 million years as well as the results of modern scientific research and modelling techniques.

6.7 CAN NUCLEAR POWER MEET THE NEEDS OF THE DEVELOPING WORLD?

In the next two decades there is little expectation that nuclear power will extend far beyond those countries who already have or plan to pursue the technology. In the developing world much of the energy demand growth is likely to be in fuels for heating and transport. In these countries growth in electricity demand will accompany urbanization and the establishment of an industrial and commercial base. Initially this will be met almost exclusively by burning cheaply produced indigenous fossil fuels and, where available, the development of particularly favourable hydroelectric opportunities. These will be the most affordable in the short term. Such developments will tend to concentrate power production close to the primary energy source - coal - and will therefore cause the establishment of distribution grid systems for the transfer of bulk electricity to the main centres of demand. It is the establishment of a grid system that is the key to bringing the benefits of nuclear power to a much wider population.

A trans-border grid system will make it possible for a country to establish interconnectors to neighbouring countries over which to export or import electrical energy, including nuclear energy. There are countries which already export their nuclear energy, for example France currently exports around 15% of its nuclear energy to all its neighbours including Britain.

With an interconnected system, several countries can join together to own jointly a nuclear installation that is beyond the means or needs of any one or two of them individually. Daya Bay, built in the People's Republic of China and serving Hong Kong, is just one example. Eventually, each country will seek to reach the point where it can invest inside its own territory in the nuclear plant required to meet internal power requirements. System interconnections will continue to be extremely valuable but the economics of energy transport are such that it should always be preferable to transport nuclear fuel in preference to electricity. It has to be the aim of the nuclear industry and concerned governments to do everything that is necessary to make this both affordable and generally acceptable.

6.8 BARRIERS TO THE FUTURE EXPLOITATION OF NUCLEAR POWER

The most direct and serious threat to the successful use of nuclear power in providing the energy needs of future generations is the ongoing problem of public perception and political will. Public and political acceptance go together in that each has an influence on the other, and politicians, in the end, have absolute control over the market environment through their treatment of regulation and the treatment of externalities through fiscal and subsidy policies.

In the eyes of the public, the nuclear industry has suffered from many things. It is identified with the horrors of nuclear weapons and there is a widespread fear of radiation and its potential for causing cancer. There has been technical arrogance, secrecy and the damaging impact of internecine battles between competing groups of nuclear proponents, commercial groupings and researchers. The industry also has a legacy of apparently unfulfilled promises – on safety, project time and cost and overall economics – and, with the benefit of hindsight, a reputation for grossly overstating the case for need.

There is an understandable lack of trust in nuclear energy and those who promote it, and fertile ground for all those who seek to heighten these perceptions in their campaigns to prevent its use in future.

6.8.1 Establishing a track record

The future focus must be on establishing the right track record, providing good reasons why nuclear power should be wanted, not just needed. Public trust and political will are essential if nuclear power is to be allowed to achieve what is expected. More has to be done by the industry to recover public trust. The industry is working hard to deliver what it promises and is focusing on achieving much greater agreement between those specialists that influence its standing. The author takes some comfort from the evidence that approval for nuclear power is in direct proportion to the level of direct experience and understanding. Openness in everything and two-way communication are the keys, listening to others' concerns and responding positively.

We must gain wide ownership of solutions to long-standing areas of challenge – health risks, accident risks, decommissioning, waste, reprocessing and proliferation issues - and seek common agreement that the benefits outweigh the costs.

6.8.2 The Case for Need - is the WEC report another scare story?

Since the Second World War there has been a history of overstatement of the need for all forms of energy. Extrapolation of the demand growth experienced in the 1950s and 1960s, especially for electricity, led to wild claims for the rate of expansion of nuclear energy. Even as recently as 1975, the expectation world-wide for thermal and fast reactors for the year 2000 was four times what will actually occur. Such claims are not the prerogative of the nuclear industry. Fuel cells, electric cars, the hydrogen economy and renewable technologies have been promoted with equal vigour.

The WEC scenarios could be looked upon as more of the same - just 'growth delayed' - but that would be unfair. Where the expectations of the 1960s and 1970s were 'profligacy sustained', the WEC now paints a picture of the lowest rate of growth that can satisfy the basic needs of the world's population. If demand falls short of this it will most probably be because many continue to be energy poor and suffer from disease and malnutrition. If there is any morality in world politics, this must lead nations to strive to achieve the energy, environmental and conservation targets the WEC has set out.

6.8.3 Is the environmental imperative overstated?

The truly independent experts openly admit that although acid emissions are damaging it is not certain that there is a problem over global warming and, even if there is, the experts are not sure of the consequences. But they are sure that the impact of man's use of fossil fuels on the environment is measurable and significant. The concentration of CO_2 in the atmosphere has increased by 26% since the industrial revolution. Around 23×10^9 tonnes of CO_2 are released into the environment per annum and this could rise to 37×10^9 million tonnes over the next three decades.

If the predictions for the effects of global warming are anything like correct and do occur, the world will face circumstances far more threatening to life on this planet than have ever been experienced before. The nuclear industry takes comfort from the WEC analysis that says that if everything non-nuclear succeeded beyond reasonable expectation, it would still not be enough to satisfy the precautionary principle.

6.8.4 Is sustainablity really a problem for today?

There is a view that markets will provide all the answers to problems and one is wrong to presume that tomorrow's solutions are limited to what is available today. However, the world economy has moved historically from sole reliance on the sun – still the dominant energy provider – through less diffuse resources such as the wind, to the real power behind the industrial revolution, coal, and on to oil, gas and uranium.

Sustainability at the highest possible level of economic activity must be the goal. Energy provision must be less and less dependent on finite fossil resources – surely the world's capital stock of these resources must be husbanded. Perhaps we should take a lesson from business in this respect. Businesses seek to sustain themselves and usually to grow. They know that the capital assets that they build have finite lives, and therefore good business practice is to make financial provision to be able to replace these assets. How can this principle be applied to the depletion of our 'free' but finite reserves of oil and gas? Is it sufficient, having located one reservoir, pierced it with a well and drained it to exhaustion, that the only provision necessary is to go and find another and deplete that too? What happens when the last has been found and exploited? Should there not be some reserve of capital that would at least replace everything that has been consumed? Perhaps the prices of gas and oil should include some provision for their eventual replacement by converting coal, and so on?

6.8.5 Is the developing market environment appropriate to meeting the identified needs?

In centrally planned economies and those where energy supply policy is determined by governments, the market is not an issue. Economies with insufficient access to low cost fossil fuels will be strongly motivated to invest in nuclear plant and its future development. Thus France, the CIS and a number of the Pacific Rim countries have little choice but to pursue nuclear. Where there is no shortage of oil, gas and cheap coal, the market incentives for nuclear and renewables are non-existent and it is only under some form of government or regulatory intervention in the market that either will proceed. Thus comes about the current standstill in the United States, the United Kingdom and Germany.

There is an additional problem in countries where the current overall wealth of the nation is able to afford strong organized opposition, concentrating its efforts on building and reinforcing an understandable public suspicion of technology and dread of things nuclear. Such opposition is now even extending to renewables, like wind, because their local environmental impact is resented so strongly and not because their economics are so poor.

It is entirely within the powers of governments to influence markets – one of the priority recommendations of the WEC – but they do not always have the knowledge or courage to generate the necessary political will to act in the ways and areas that need it.

6.9 WHAT ARE THE ALTERNATIVES?

The author accepts the logic and morality of the WEC's analysis of future energy demand and has demonstrated that the nuclear industry is capable within existing knowledge to deliver all that is required of it.

Light-water reactors currently supply over three quarters of the world's nuclear energy and will remain the dominant reactor system for the next few decades. However, the next generation of light-water reactors will be of the advanced design. These future designs are already well developed. In the United Kingdom, the construction of one of the world's most advanced pressurized water reactors, Sizewell B, is nearing completion. Other designs include Framatone's N4, and Mitsubishi's Standard 4-loop. While these future designs are similar to those in general service today they will include further design improvements and safety enhancements as well as having increased outputs at around 1500 MW_e.

Is there a better system waiting in the wings? Is it better to wait for fusion? Can we do without nuclear at all and simply rely on renewables and conservation, as some have proposed?

6.9.1 Future light-water reactors

The most radical developments of light-water reactor design are towards smaller integrally safe designs, of which the Westinghouse AP-600 is an example. The attraction is a move to simpler designs with increased passive safety systems that rely on gravity, natural circulation and stored energy, for accident prevention and mitigation. At around 600 MW_e they represent a better match to the market in countries without a large centralized electricity industry. In the short term, market match is not a problem because those exploiting nuclear power see it as a strategic matter and structure their industry to accommodate the larger, more economic plant. It is possible that these designs will be competitive with the best evolutionary designs but the scale of nuclear contribution that is being sought would challenge the wisdom of preferring smaller units with the attendant doubling or trebling of the number of reactors required.

6.9.2 Fast reactor

In most realistic energy growth models, fast reactors may be required around the middle of the 21st century. Anyone who has visited the 1200 MW fast reactor at Creys Melville, near Lyon, will recognize a technology that is scientifically and technically proven at a commercial scale. While this plant is not considered able to compete with current light-water reactor technology, it is easy to imagine that this would be possible with limited further development and a commitment to a sufficiently large programme to bring the full benefits of scale the French have already demonstrated in their PWR programme. A commercial fast reactor is feasible today.

However, if fast reactors are to be a major part of our energy future, they must be developed further to ensure that the commercial risks are clearly understood and minimized well before there is a major commitment. This cannot be achieved without further investment and there is even the risk that our existing fast reactor technology will be lost. Whatever civil servants think, engineering expertise cannot be sustained in a book.

The UK Government has expressed the view that the civil nuclear industry should shoulder the financial burden of this long-term strategic research. This is not possible in the competitive commercial environment in which the UK nuclear utilities now operate. In the future we will have to rely on the French and Japanese to sustain and supply this technology.

6.9.3 Fusion

Fission and fusion were discovered within a few years of each other, and while fission power is now a maturing industry, fusion has yet to demonstrate even scientific feasibility. When the author was a young engineer success was 40 years away. Forty years on it is clear that it is still 40 years away. There has been enormous progress in understanding the physics of tokamak fusion over the past 40 years, and yet this first stage is still incomplete. The science is proving exceptionally hard to master and the practical engineering will be harder still.

There are two fundamental engineering issues. The first stems from the geometry of a tokamak. To remove the heat from the fusion reaction requires the handling of very high heat fluxes, some 10 times greater than the maximum found in a modern PWR core, or the production of lower flux densities. This can only be achieved by significantly increasing the size of the fusion reactor.

The second issue stems from the nature of fusion energy, 80% of which appears in the form of fast neutrons, which damage structural materials, degrading their mechanical properties causing distortion and swelling. Even if new super-materials could be developed which out-perform today's the first wall of the structure would still have to be replaced every few years.

A fusion reactor would be about 10 times bigger, in terms of real reactor plant, than a PWR of similar output. Even if the cost per unit volume were the same - which is unlikely since fusion technology is far more complex - this would imply capital costs for fusion plant around 10 times higher than a PWR. Paper studies have promised costs closer than this, but turning paper studies into hard engineering is another matter. In addition, the need for regular replacement of the torus walls and the risk of high rates of tripping from the hugely complex control systems, threaten low plant availability compared with today's reactors. This would make the economics of fusion plant even worse.

The next planned step in harnessing fusion is the International Tokamak Experimental Reactor (ITER), which will have a total birth to death cost exceeding US$10 billion. This is a massive, even unreasonable, cost to contemplate at this stage when there are still too many unanswered questions. For example, should the next step be a tokamak or should it be inertial confinement? The physics of inertial confinement seems to have caught up with magnetic systems, but the step to power production looks even harder.

Fusion is a research field with some promising lines and many culs-de-sac. It should be treated as such and funded accordingly until the best way forward is much clearer. I see no serious prospect of fusion contributing to commercial energy supply in the timespan of the 21st century.

6.9.4 'New' renewables

It is difficult for the author to challenge the potential of renewables without arousing the obvious charge of self-interest, yet he is no evangelist for nuclear power. The author would not be averse to playing a major role in the commercial expansion of renewable generating systems, but does have serious reservations about them that do not seem to form part of the present debate. Reference has already been made to the high capital cost associated with all the 'new' renewables, and the author wonders if they really could provide more than today's total energy output from all fuels within 50 to 100 years.

The author also wonders about their sustainability. All renewable resources are, by definition, sustainable. But the plant to extract the power requires the investment of energy in its construction, operation and eventual replacement. Recent work by Keele University (Proops, Gay and Speck, 1994) has given an insight into the scale of direct and indirect resource requirements. Table 6.5 shows their results comparing the time it takes for

Table 6.5 Time (days) to generate the energy used in construction and decommissioning

Type of plant	*No. of days*
Gas	45
Coal	66
Sizewell C	160
Tide	391
Wave	647
Wind	885
Solar	1948

the energy generated to repay the energy consumed in construction and decommissioning. Other parts of their analysis suggest that the indirect energy and environmental costs associated with their operation are also surprisingly large. It may be that a major renewables programme based on any of today's technologies will not be possible without the continued support of a fossil or nuclear resource base. This should be examined further before too much reliance is placed on renewables.

6.10 CONCLUSIONS

Compared to earlier times, such as the Club of Rome, the likelihood is that the demands on fossil fuels and nuclear generation are under- rather than over-estimated – especially in the first part of the 21st century. A major new dimension is the shift in demand from the developed to the developing nations of the world.

Sustainable growth of both the world's economies and their energy supplies is a necessary aim and substantial progress must be made within the next few decades. Above all, this growth must be achieved without risking catastrophic environmental damage.

There are major short-term challenges.

1. Energy intensity must be reduced – continued relative growth of the electricity sector is the key.
2. Energy production – direct or via electricity – must shift more to nuclear and, if possible, renewables.
3. Developments must meet the needs of the developing world – in terms of technology and supply.
4. Diversity of supply and its widespread availability are essential also to avoid the risk of future international conflicts over the maldistribution of resources.

Renewables may satisfy these objectives – but cannot be relied upon technically to deliver what is expected of them. Their capital costs are high and it is by no means clear that they will be affordable at the scale being sought. Their sustainability has so far been assumed without challenge – the author suggests that this assumption should be examined with great care.

By contrast, nuclear power is already established and technically capable of meeting its challenge and of filling the gap which shortfalls elsewhere may cause. However, nuclear also requires support from the public and governments if it is to be able to compete with fossil fuels in many of the industrial economies that hold the key to nuclear technology and its development. Wherever the industrial capability to design and manufacture nuclear plant is allowed to atrophy, it will take several decades to re-establish and, therefore, would not be available to meet a rapid rise in demand that any regional decline in oil and gas supplies may cause.

It is not the responsibility of the nuclear industry across the world to secure the energy and environmental benefits we believe the world requires without there being the appropriate incentives and rewards in the market. Only governments have the responsibility to make such decisions on behalf of the people they are elected to serve and represent. If governments do not accept the WEC advice and act quickly on the recommendations, it will rapidly become difficult for the world's nuclear industry to respond on the scale and in the time required. All our energy and environmental futures are in their hands.

ACKNOWLEDGEMENTS

The author thanks his many colleagues in Nuclear Electric, in particular John Collier and Ian Glendenning for their valuable contributions to this chapter, and Gail Brannick for her careful compilation of the information.

REFERENCES

Flowers, Sir B. (1976) *Royal Commission on Environmental Pollution - Sixth Report*. Cmnd. 6618, HMSO, London.

Proops, J.L.R., Gay, P.W. and Speck, S. (1994) *The Lifetime Resource and Pollution Implications of Various Types of electricity Generation: An Input - Output Analysis*. Department of Economics, Keele University Report.

World Energy Council (1993) *Energy for Tomorrow's World - the Realities, the Real Options and the Agenda for Achievement*. Kogan Page, London.

7

The future for renewables

K.J. Brown

7.1 INTRODUCTION

Renewable sources of energy have enjoyed a considerable renaissance in the late 20th century. In the developing world, wind, water, sun, waste and crops have begun to make a significant contribution to meeting heat and power needs. Developing countries, too, are extending their traditional reliance on renewable energy sources.

In the third millennium, the renewables are likely to play an increasing role at home and abroad, largely because of their ability to help combat both global warming and other environmental problems causing concern to the international community. A recent UN study concluded that, in certain circumstances, renewable energy could account for as much as 60% of the world's market by the mid-21st century.

Realizing that potential, however, depends on a number of factors, such as stimulating a market, addressing the barriers to the renewables' deployment and raising awareness of the benefits and opportunities these clean energy sources offer. Only then will optimistic projections of the renewables' future prove justified.

7.2 RENEWABLES IN PROSPECT AND RETROSPECT

Like the end of a year or decade, the end of a millennium represents a watershed, an opportunity not only to look back and take stock, but also to look forward. If we subject renewable energy – the term that covers the wide range of naturally occurring and replenishable energy sources (Table 7.1) – to this process of review and prediction, a fascinating picture emerges.

A retrospective of the last 1000 years reveals a neatly cyclic picture. At the beginning of the second millennium, energy sources such as wind, water and wood were unrivalled as providers of heat and basic mechanical power.

Table 7.1 Principal types of renewable energy

• Wind power
• Hydro power
• Wave power
• Tidal power
• Municipal wastes
• Industrial wastes
• Landfill gas
• Energy crops
• Agricultural and forestry wastes
• Passive solar design
• Active solar
• Photovoltaics
• Geothermal heat

The Domesday Book, compiled in the late 11th century, records the existence of 5000 waterwheels in England alone. In more recent times, however, new energy-providers such as coal, gas and oil, and latterly nuclear power, displaced those traditional energy sources. This was especially the case where countries became industrialized, although it should be noted that the role of renewables was not eroded to the same extent in non- or under-industrialized parts of the world.

In the 20th century, and in particular since the oil crises of the 1970s, the renewables have enjoyed a genuine global revival. A variety of factors have brought about this renaissance – for example, the renewables' capacity to aid economic growth in underdeveloped regions beyond the reach of centralized power generation and distribution systems. But the primary catalysts have been, firstly, the need to ensure diversity and security of energy supplies and, secondly, the need to protect the environment by offsetting problems, such as global warming, associated with fossil fuels. The net result is that, today, traditional renewable energy sources such as wind, hydro and biomass, and newer ones such as waste, are meeting about one fifth of the world's growing energy needs.

The fundamental catalysts that have driven the renewables' recent development – or more correctly their *re*-development – (i.e. environmental benefits and diversity of supply) will almost certainly continue to be an influence in the future. So, as we look ahead to the third millennium, the overwhelming – and generally welcomed – probability is that renewable energy will play an increasing role in the energy mix. What is open to question, though, is the exact extent of the renewables' potential and how far that potential will actually be realized.

Naturally, these uncertainties have led to significant effort being expended in assessing the likely rate of the renewables' deployment. Indeed, predicting the renewables' future has become almost an industry in itself, with considerable intellectual resources and powerful computer models dedicated to the pursuit of authoritative answers. In general, the resulting assessments indicate a potentially bright future in the United Kingdom, in Europe and world-wide.

But those predictions bring a danger with them. It could easily be assumed that a bright future for renewables is guaranteed, merely because such a future is possible; it could be taken for granted that favourable projections will somehow inevitably come true. In reality, however, it is not the predictions themselves that are crucial, but rather the assumptions that lie behind them. Those assumptions indicate what must happen - in terms of the necessary level of government and industry commitment to the renewables' development, and the extent to which past lessons about incentives, barriers and attitudes are taken on board - if optimistic scenarios are to prove justified. For, in the final analysis, the future of renewables will not be determined in 20 or 50 years' time, but largely by the choices and decisions that are made now and in the next few years.

This chapter will look at the renewables' prospects at home and abroad, ask whether estimates of their impact into the next millennium are realistic, and distil the key factors that their impact depends upon.

7.3 THE WORLD PERSPECTIVE - TWO RECENT STUDIES

In 1993, two influential reports were published addressing the future world-wide role for renewable energy. The first of these, *Renewable Energy - Sources for Fuels and Electricity*, commissioned by the United Nations, looks well into the 21st century (Johansson *et al.*, 1993). Against a background of a world economy in expansion, it draws the broad conclusion that, in conjunction with energy efficiency measures, 'given adequate support, renewable energy technologies can meet much of the growing demand at prices lower than those usually forecast for conventional energy'.

More precisely, under a 'renewable-energy intensive global energy scenario', the report suggests that by 2050 the renewables could account for as much as 60% of the world's electricity market and 40% of the market for direct-use fuels. This would help reduce carbon dioxide emissions to 75% of their 1985 levels. These figures - primarily bolstered by an assumed increase in the deployment of hydropower, wind energy, solar energy, geothermal energy and biomass (principally energy crops grown on land that is either deforested, degraded or excess to crop production) - are impressive enough, but the authors emphasize that they by no means represent a ceiling to the renewables' potential: 'higher levels might be pursued, for example, if society should seek greater reductions in CO_2 emissions'.

A number of assumptions underpin this scenario – for instance, that declining costs and pressures to reduce damaging emissions will produce an investment climate favourable to renewable energy. Indeed, the conclusions are basically predicated on governments, moved by environmental imperatives and the need for secure energy supplies, intervening to subsidize renewables and remove subsidies currently enjoyed by conventional energy options, recasting their agricultural policies, and increasing support for renewables research, development and dissemination.

To highlight the lessons that can be learnt from past experience, the report outlines a number of real-life examples of renewables' development. Brazil's alternative transport fuel programme is a particularly interesting one. In the 1970s, through a battery of measures such as tax and price incentives, the Brazilian government launched an initiative to promote the use of ethanol as a transport fuel. The net result has been that about 60% of all cars in Brazil now use a fuel comprising 100% ethanol. In recent years, however, as world oil prices have fallen, ethanol has become uneconomic and Brazil is under pressure to run the programme down. Fewer ethanol-only cars are being sold, but those running on petrol/alcohol blends are becoming popular instead. Indeed, the report argues that sugarcane-derived alcohol can be economic at today's oil prices and concludes that Brazil is well placed to exploit the opportunities that present themselves to renewables in the transport sector.

California provides a second example. This state can boast a very well developed wind and solar energy resource (i.e. about 2000 MW_e of capacity either operating or under construction), principally as a result of tax credits and guaranteed markets for renewables in force during the 1980s. A number of poorly designed wind turbines were an unfortunate by-product, however; in addition, the company responsible for building California's solar thermal-electric capacity went bankrupt as soon as subsidies were (somewhat abruptly) curtailed. 'Programs must send consistent signals to investors', the report concludes. '[Subsidies] should be reduced according to a clear and predictable schedule.'

Finally, Denmark's wind energy programme provides a genuine success story. Launched in 1979, it involved the Danish government giving grants for part of turbines' purchase and installation costs. It was stipulated well in advance that the subsidy would be phased out by 1989, although wind energy in Denmark continues to benefit from certain other concessions; in addition, the subsidy was only applicable to turbines tested and approved by the Risø National Laboratory. The result has been a vibrant and rapidly expanding Danish wind industry, with an internationally successful manufacturing base. Denmark now produces some 2.5% of its electricity from wind power, with nearly 3000 grid-integrated turbines in place.

This 1000-page report prompts a number of conclusions. The most fundamental is that, while the renewables have huge potential, its robust realization will be crucially dependent on policy and investment decisions. Nor are these decisions only a matter for governments, deciding whether to set up a framework for the renewables' exploitation, and for developers, looking to take technologies from the drawing-board and deploy them in the real world. It also to a significant extent concerns individual consumers, who may choose to embrace and welcome renewables, for environmental or economic reasons - as in Brazil - or draw a negative conclusion about renewables as viable energy options.

The second report, the World Energy Council's *Renewable Energy Resources: Opportunities and Constraints 1990–2020*, provides another optimistic, if slightly more conservative, view of the renewables' future (WEC, 1993). It describes how, today, 18% of the total energy used in the world is derived from renewables, primarily hydro power and biomass. Under a 'current policies' scenario, which assumes falling costs for renewable energy technologies, the study argues that this figure could 'relatively easily' rise to 21% by the year 2020. Given strong international support at government level, it says, this figure might even reach 30%, with much of the impact felt in developing countries, where the renewables could help combat problems such as the migration of rural populations to cities.

The report states, however, that such an expansion in use depends on a number of factors, such as changes in economic policies and in the way decisions on energy issues are made, an increase in funds for research and development, and the removal of subsidies enjoyed by conventional energy sources. Furthermore, the report underlines some of the barriers which could hamper deployment, such as 'existing investments and interests in fossil fuel and nuclear energy provision and use' and 'short-term planning based on current economic decision-making patterns'. The study also notes that, despite their environmental advantages at a global level, technologies such as wind and hydropower 'can have significantly adverse local impacts'.

To conclude this review of two important and comprehensive studies, it should be re-emphasized that we must not become obsessed with projections and predictions about the role of renewable energy. If a growing contribution is to materialize, a whole range of barriers and issues will have to be addressed. Indeed, we as a world, as nation states, and as individuals will first have to make the most fundamental decisions about what we really want from renewable energy.

7.4 THE EUROPEAN ROLE FOR RENEWABLES

The Commission of the European Communities (CEC) has already recognized the need to make decisions of this kind. Indeed, its intention is to ensure a much greater role for renewables across the Member States, thereby helping to limit both 'greenhouse' gas emissions and European dependence on imported energy. The introduction of a Community-wide energy tax on users of carbon-rich fuels has already been the subject of considerable discussion around Europe; however, even if such a tax should come into force, the CEC recognizes that it would not in itself be sufficient to prompt a substantial boost in the deployment of renewables.

As a result, the Commission is now launching a new Alternative Energy Programme, known as ALTENER. ALTENER's central objective is to double the renewables' contribution to total energy consumption in the European Community from 4% to 8% by the year 2005, to include the trebling of electricity production from renewables sources. Primarily, this 'hard task, but . . . achievable goal' will be pursued through an expansion in biomass, biofuels, wind power and hydro power (CEC, 1992).

ALTENER will operate across a broad spectrum of spheres. Promotion of the internal market for renewables will be furthered by seeking to establish a Community-wide legal framework conducive to the renewables' deployment, the lack of which has been identified as a major obstacle to development in the past. A raft of financial and economic measures is envisaged, such as the establishment of a network of financiers interested in project financing renewable energy schemes. Training and information activities will aim to raise industrial awareness of the renewables' potential.

The ALTENER programme, then, constitutes a recognition of the need for positive immediate action if the use of renewable energy is to grow significantly. Indeed, it reveals the CEC's awareness that national programmes of grants, subsidies and tax and other incentives already in place around Europe, for all their merits, will not be enough to secure the renewables' medium- and long-term future. Thus ALTENER aims to 'co-ordinate national initiatives and fix a common goal'. Yet even so, there remains a recognition that the renewables' European future does not lie entirely in the hands of Europe's decision makers. If the 8% target figure is to prove realistic, the CEC concedes, 'oil prices would have to rise to entice purchasers to buy renewable energy sources'.

7.5 THE UNITED KINGDOM AND RENEWABLES – CURRENT STATUS

In the United Kingdom, too, decisions on the future of renewable energy have already been made, primarily by the Government but also, increasingly, by developers and investors. The renewables currently meet about 1% of the United Kingdom's total energy requirements, and their recent and current development has been underpinned by the Government's New and Renewable Energy Programme. Now the responsibility of the Department of Trade and Industry, which absorbed most of the former Department of Energy, the programme has been managed by ETSU since its inception in 1974. Its central aim is to stimulate the use of new and renewable energy technologies wherever they have prospects of being economically attractive and environmentally acceptable. Over £300 million (in 1993 money) has already been invested in developing the technologies to the point where a number are now making a significant contribution to UK energy supplies.

Perhaps the biggest single boost to the renewables' deployment in the United Kingdom, however, has been the Non-Fossil Fuel Obligation (NFFO) introduced at the time of the privatisation of the electricity supply industry in 1989. The NFFO requires the 12 Regional Electricity Companies of England and Wales to derive specified amounts of their power from renewables sources, with premium prices payable for electricity generated funded by a levy on sales of electricity to final customers. The first and second Orders, which came into effect in 1990 and 1991, awarded contracts to a total of 197 renewables schemes for a total nominal capacity of 624.5 MW (see Table 7.2). A third round is scheduled for later in 1994, along with broadly parallel Obligations for Scotland and Northern Ireland. Further Orders are in prospect for 1996 and 1998. As at the end of December 1993, 134 NFFO schemes were operational, with a total capacity of 251.2 MW DNC.

Table 7.2 NFFO Renewables tranches 1 and 2

Technology/ energy source	*Number of contracts*	*Size of contracts (MW)*	*Contracted capacity (MW DNC)*	*Contract price p/kWh*
First tranche, 1990				
Landfill gas	25	0.18–4.8	35.5	
Sewage gas	7	0.17–3	6.5	
Waste incineration	4	1.65–27	40.6	
Hydroelectricity	26	0.04–3.8	11.8	
Wind	9	0.03–3.9	12.2	
Other	4	0.1–20	45.5	
First tranche totals	75		152.1 MW	
Second tranche 1991				
Landfill gas	28	0.4–5.0	48.5	5.7
Sewage gas	19	0.3–7.8	26.9	5.9
Municipal and general Industrial waste	10	4.1–103	271.5	6.5
Hydroelectricity	12	0.02–0.6	10.9	6.0
Wind	49	0.1–7.4	84.4	11.0
Other waste combustion	4	1.0–12.5	30.2	5.9
Second tranche totals	122		472.4 MW	
First and second tranche totals	197		624.5 MW	

The NFFO represents the principal means by which the Government aims to reach the broad objective set for renewables development in the next few years – 1500 MW of new renewables generating capacity by the year 2000 (i.e. capacity additional to the 1.5 GW of primarily large-scale hydro plant that pre-existed and was not included in the NFFO, and the 100 MW of pre-existing renewables plant included in NFFO 1 or NFFO 2). That aim represents one dimension to the Government's response to the threat of global warming and the need to offset carbon dioxide emissions to the atmosphere.

Naturally, most of the envisaged new capacity will come from technologies that are at or close to technical maturity. Hydro power is one example. The UK's first large-scale hydro scheme was built in Scotland in 1896 and there are now numerous projects around the country, ranging from installations producing less than 1 kW to those generating more than 100 MW. Although new large-scale schemes are unlikely to be built, for environmental reasons, 38 small-scale schemes (i.e. with an individual generating capacity of up to 5 MW) are already contracted to supply electricity to the Grid under the NFFO, with further such sites also offering potential for development. Overall, hydropower today accounts for about 2% of the UK's electricity supply.

The United Kingdom has the largest wind resource in Europe. Although the modern exploitation of wind power does not have as long a pedigree as hydropower, there are already well over 300 wind turbines in operation around the country, providing electricity not only for the Grid but also for local use in Grid-remote situations. A total rated capacity of over 125 MW has already been commissioned or is currently under construction – almost entirely as a result of the NFFO.

The United Kingdom is currently a world leader in the exploitation of landfill gas as an energy source. This gas, produced by the microbial breakdown of waste deposited in landfill sites, can be used to produce power for the Grid or heat for boilers, kilns and furnaces. Over 60 landfill gas-to-energy projects have already been commissioned. The United Kingdom, which disposes of 90% of its waste in landfill sites, is the largest user of landfill gas as an energy resource in the European Community.

Also in the waste area, a range of waste materials, including both municipal solid waste and some unlikely sources such as chicken droppings and scrap tyres, are providing energy through combustion – where appropriate, as part of an integrated waste strategy involving recycling. A power station at Eye in Suffolk is using chicken litter to generate 12.5 MW of electricity for export to the Grid, while a plant at Wolverhampton is producing 25 MW of power for the Grid through the incineration of 94 000 tonnes of waste tyres a year. In both cases, state-of-the-art technology ensures compliance with all environmental legislation regarding emission limits. Wood, a more conventional, indeed traditional, energy source that can be burned efficiently and cleanly, also has significant potential. Growing coppiced willow and poplar as an energy crop could offer farmers an attractive, 'green' means of farm diversification and of making the most of land set aside from food production. Its viability is still being assessed, primarily through the DTI's Farm Wood Fuel and Energy Project, which aims to develop growing centres and stimulate markets for arable energy coppice.

Solar energy, too, is important. Nearly all buildings make some use of it. Furthermore, passive solar design, whereby the form, fabric and orientation of a building are used to capture the sun's energy for heating, lighting and ventilation, is used in a number of buildings in the United Kingdom. Active solar heating, which uses solar collectors to transform solar radiation into low temperature heat, is mostly used to provide domestic hot water or to heat swimming pools. The technology has been widely applied in the United Kingdom, and there are over 40 000 installations in place. Finally, photovoltaics, where solar radiation is converted directly into electricity, is still in its infancy and used only on a very small scale in the United Kingdom.

A range of other renewables resources are making either an extremely small contribution to energy supplies (i.e. wave energy, geothermal heat) or are yet to be deployed in the United Kingdom (e.g. tidal power, offshore wind). These limitations are primarily due to either the projected cost of resulting electricity being uneconomic (tidal power, offshore wind and wave energy) or a low level of UK resource and technical difficulties in harnessing it (geothermal heat).

In general, then, the current UK picture is clear. The renewables are beginning to make their mark. The Government is setting a framework for deployment. Developers and investors are beginning to come forward to capitalize on the commercial opportunities. But what of the future? Can this progress be sustained?

7.6 THE UNITED KINGDOM AND RENEWABLES – PROJECTIONS

The future path for renewables was addressed by the Renewable Energy Advisory Group (REAG) in its *Report to the President of the Board of Trade* in November 1992. Comprising 12 members and including independent industrialists, scientists and economists, REAG was set up in mid-1991 to spearhead the Government's current review of renewable energy strategy. In the course of its work, it received over 1000 written statements, heard 18 oral presentations, and visited 17 organizations and installations (HMSO, 1992).

Drawing on interim findings from ETSU's assessment of renewable energy sources, undertaken on behalf of the DTI, the REAG Report describes the theoretical accessible UK potential for renewables (i.e. the amount which it would be technically possible to derive, taking into account only basic constraints) as in the order of 1100 TWh/year of electricity generation, at less than 10 p/kWh (1991 prices); this equates to three to four times the current UK supply of around 300 TWh/year. In reality, it would not be possible to exploit this level of resource; as REAG notes, exploitation on such a scale would necessitate large areas of the United Kingdom being given over to renewable energy projects.

To obtain estimates of the theoretically possible future economic contribution of renewables, therefore, it was necessary to put theoretical resource availabilities into the context of possible future UK energy supply markets. Preliminary results from ETSU indicated that about 10–60 TWh/year and 15–130 TWh/year might be economic by 2005 and 2025 respectively (see Figure 7.1). The Report notes that 'achieving the top end of these ranges assumes high future electricity prices and low discount rates. The Group considers that such contributions are unlikely to be delivered, but that renewables could supply as much as 20% of current UK electricity demand by 2025 under severe pressures of need and economics.'

Figures 7.2 and 7.3 present preliminary supply curves, included in the Report, for the electricity-producing renewables in 2005 and 2025, for a representative commercial discount rate of 15%. While noting that such estimated contributions were of considerable interest, however, REAG concluded that 'they were inevitably dependent on assumptions made about the perceived costs of energy (including environmental costs), values put on environmental attributes (including scenery free from visual intrusion), and the social acceptance of current waste disposal patterns'. Building on this, the Report states that, in REAG's view, 'a major factor in determining the size of the estimated contribution will be the extent to which general energy prices rise as fossil fuel sources bear the costs of the environmental damage with which they are associated'. In addition, REAG noted a wide variety

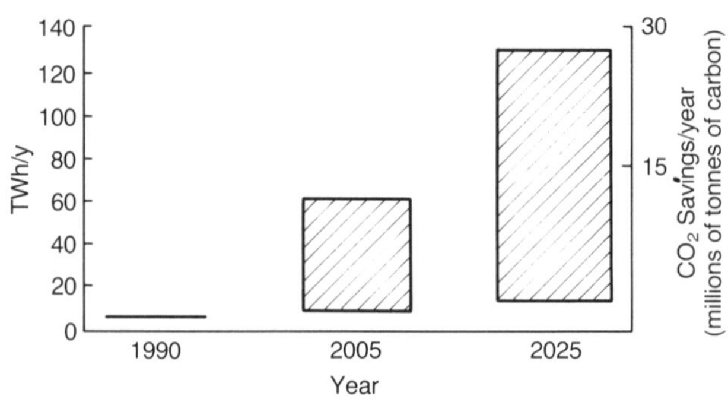

Figure 7.1

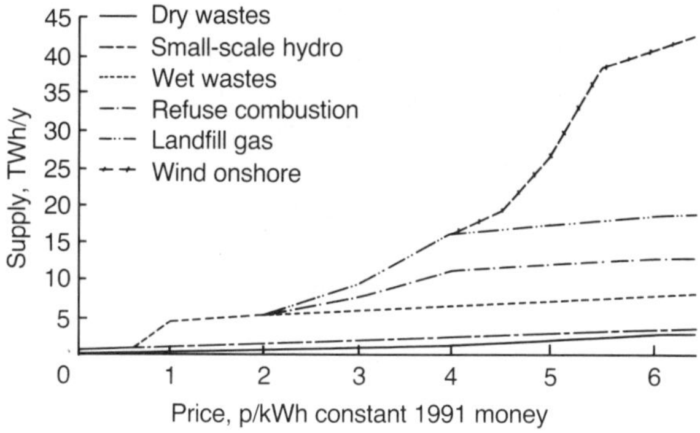

Figure 7.2

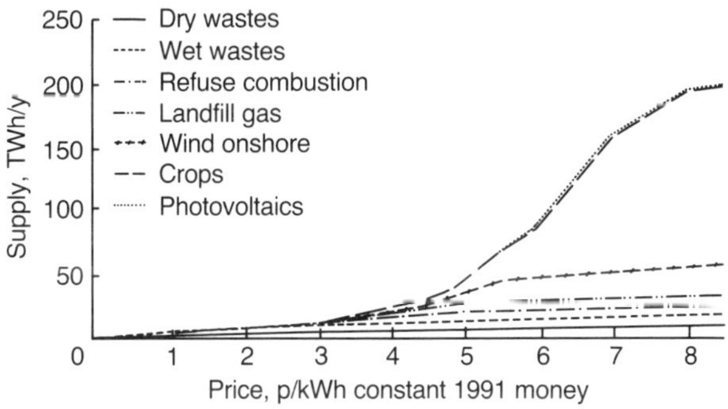

Figure 7.3

of other factors which necessitate a degree of uncertainty over the renewables' future contribution: for example, 'inertia and institutional barriers', 'perceptions of risk' and 'the absence of any historic market in renewables, which has resulted in legislative and administrative frameworks becoming established without particular regard to the needs of renewable energy'.

7.7 THE UNITED KINGDOM AND RENEWABLES - REMAINING BARRIERS

The message is clear: if we in the United Kingdom are to derive maximum benefit from renewable energy, a number of remaining barriers need to be tackled, and a number of choices need to be made. Perhaps the central underlying problem facing the renewables derives from the comparative novelty of most of the technologies in their modern form. Experience of their use as energy providers is limited and the energy industry therefore needs to be fully convinced that renewables are viable and that returns on investment can be competitive. In the current economic climate especially, real and indeed perceived risks can be a major deterrent. The NFFO provides a platform for the commercial demonstration of maturing technologies and the DTI is also highly active in raising awareness and providing the market place with information as an aid to investment decision making.

As the REAG Report points out, the relative 'newness' of renewables also means that a legislative and administrative framework has emerged for the energy market place that was not designed to accommodate the particular needs of renewable energy technologies. Potential developers wishing to deploy renewables have encountered difficulties with the interpretation and application of regulations in areas such as environmental protection, pollution control, waste management and disposal, use of conservation areas, water abstraction and planning.

In particular, local authorities' unfamiliarity with renewable energy - in many cases symptomatic of unfamiliarity with renewables among local communities as a whole - has been a factor in the significant problems encountered by some renewables projects when seeking planning permission. Renewables have a 'green' image and are undoubtedly beneficial to the global environment, but projects can nevertheless have a significant local environmental impact. In some instances, this can prompt antipathy to individual schemes - even among those who would consider themselves 'green' in other circumstances. Wind power is the classic case. Although emitting no carbon dioxide or other damaging emissions, wind farms do have a physical presence, impacting the landscapes where they are sited; furthermore, some people living close to developments have expressed concerns about the level of noise produced by turning turbines.

A recent DTI survey undertook a detailed examination of local opinion surrounding the UK's first wind farm, which became operational at Delabole in Cornwall towards the end of 1991 (ETSU, 1993). Interviews were conducted with a sample of around 300 people living within 2 km of the site. Before the wind farm was built, 31% of the interviewees said they were not in favour of the development, while 40% approved. The sample was then reinterviewed after the farm had been operating for six months. Views had changed markedly in the intervening period. In the second survey, 85% of interviewees said they approved of wind power. These results suggest that anticipation of the Delabole wind farm produced anxieties in some people which were successfully countered only by actual experience of the farm in operation. But Delabole, of course, represents just one wind farm site and each development is unique. The overall UK picture suggests that the planning issue will continue to be a major factor in determining the extent of wind power's deployment in this country.

Recognizing understandable concerns about renewable energy, the Department of the Environment published Planning Policy Guidance Note no. 22, which emphasizes the need for planning authorities to balance policies for the development of renewable energy with policies for the protection of the countryside. In addition, raising awareness and increasing understanding of renewable energy and its implications among the general public has formed an important part of the Government's Renewable Energy Programme as a whole.

Finance is another crucial area. Renewables projects are, typically, capital intensive and require developers to finance their schemes either from their own funds or, more usually, externally (through project financing, for example). Attracting investment to a scheme can be complex and time consuming, particularly as renewables have to compete with a whole host of other investment options, which are often perceived by lenders to be less risky.

Again, the DTI is making substantial efforts to inform the financial community about renewable energy. As a result, more and more organizations are being attracted to what they see as a growing industry of considerable promise. It is essential to put these early successes into perspective, however. Even attaining the Government's broad short- to medium-term aim of 1500 MW of new renewables capacity by the year 2000 would require a total investment of up to £3 billion. A prerequisite for investment on this scale would be the widespread recognition and pursuit by the financial community of the business opportunities the renewables can offer.

Today, then, the renewables are facing a variety of hurdles, which the Government is addressing as a matter of urgency. In highlighting finance, planning and public acceptability, this chapter has focused attention on subjective barriers. That is not to underestimate factors associated with infrastructure and hurdles of an institutional or regulatory character, referred to more briefly. But tackling these is more straightforward than the less tangible, more social barriers. If renewable energy is to flourish in the United Kingdom, and indeed in the world as a whole, planners, financiers and the public at large will all have to play their part in its development, in either an active or a passive sense. The crux is that all these people live in the real world, the world of the subjective and of the emotive, where behaviour predicted or assumed by clinically produced projections and rigorous computer forecasts does not always apply.

Consider the current position in the United Kingdom. The Government is well advanced in building a framework for the exploitation of renewable energy, to help the nation to derive the maximum benefit from economically viable and environmentally acceptable renewables technologies. The Non-Fossil Fuel Obligation is in place, providing a major incentive to potential participants in this new industry. The huge global export market for renewables-related products and services that is already developing is a further incentive impelling the United Kingdom to embrace the renewables. Finally, the broad environmental imperative, brought about by the threat of global warming, should also prompt us to welcome the arrival of these intrinsically clean energy sources.

In the final analysis, though, people will have to make their own decisions. Despite our hopes and wishes, it is impossible to second-guess what those decisions will ultimately be. What appears to an outsider to be a 'rational' choice may not be the option an individual actually takes. In the realm of the subjective, there can be no guarantees. The renewable **could** make a major contribution to meeting our energy needs, but only if we as a society really want them to.

7.8 CONCLUSION

This chapter began by considering the impact renewables will have in the next millennium, but has as its main aim attempted to draw attention away from the future and towards the present day. The temptation may be to look ahead, into the 21st century and beyond, but it is today that we need to address the issues. It is today that every one of us needs to decide what we ultimately want from our future energy industry. If that future is to encompass a growing role for renewables - and the signs are that they will make an important contribution - then we have to ensure that they can deliver. And that can only be achieved **now**, by increasing our understanding of the subject and addressing the barriers to deployment.

REFERENCES

Johansson, T.B., Kelly, J., Reddy, A.K.N. and Williams, R.H. (eds) (1993) *Renewable Energy - Sources for Fuels and Electricity*. UN Report. Island Press.

World Energy Council (1993) *Renewable Energy Resources: Opportunities and Constraints 1990–2020*. WEC Report.

Commission of the European Communities (1992) *Alternative Energy Programme (ALTENER) Background Report (ISEC/B28/92)*.

ETSU (1993) *Attitudes Towards Wind Power - A Survey of Opinion in Cornwall and Devon*.

Renewable Energy Advisory Group (1992) *Report to the President of the Board of Trade*. HMSO, London.

8

Oil and gas utilization: limits of efficiency and their impact on demand

J. Masters

8.1 INTRODUCTION

Many analysts have collated data on energy consumption for different activities in the economies of the world. They have also assessed the impact of policies to curb consumption. This chapter adds to these projections by assessing the impact of achieving the realizable limits of efficiency on world energy consumption.

Oil and gas are fossil fuels used primarily in power generation, transportation, industry, commerce and the home. Together with coal they are the staple diet on which industrialized and developing economies depend. Both fuels are in abundant supply with new discoveries currently outstripping the growth of demand. In geological time, however, the consumption of fossil fuels will occupy only a brief period of the Earth's history. Their demise will almost certainly occur during the third millennium. The rate of depletion will be accelerated by growing demand from the emerging industrial economies. If the world population is one day to enjoy the same level of amenity as the industrialized nations, then clean, renewable or inexhaustible supplies of energy will need to be developed during the lifetime of fossil fuel use.

The efficiency of energy consuming equipment can be defined as the ratio of useful energy output to the calorific value of the fuel input expressed as a percentage. Therefore, a boiler can be described as being typically 80% efficient; a gas turbine 30%. This is an expression of the efficiency of transformation of fuel into a useful form of energy such as heat or motive power. In use, however, fuel consumption is just as dependent on the application as on the efficiency of the fuel consuming component in the system. For instance, the energy consumption in a building is as dependent on the design, structure, use and location as on the thermal efficiency of its boiler. Similarly, the fuel consumption of a vehicle under differing conditions of use is a more useful measure of effectiveness than the shaft power efficiency of the engine. This chapter assesses not only the limits of equipment efficiency, but also the potential impact of improvements in system performance.

A key driver for higher efficiency at the present time is the perceived need to reduce those gaseous emissions that increase global warming. The concern is such that governments have set limits for the chief culprit, carbon dioxide. The main reason for the increase in CO_2 is thought to be the combustion of hydrocarbon fuels. However, these are essential to the well-being of the industrialized economies and to the emergence of the developing economies. If standards of living and levels of industrial activity and growth are to be maintained while curtailing the output of CO_2, then efficiency in the use of energy must increase markedly. In the shorter term, switching from fuels with a high carbon to hydrogen ratio, such as coal and oil, to those with a lower ratio, such as natural gas, can make a useful contribution and there is strong evidence of this trend around the world.

Energy consumption for the variety of human activities differs geographically. For instance, there are major differences in the levels and patterns of consumption between the United States, Japan and Western Europe, and the former USSR and the developing and undeveloped economies. To obtain the potential impact of increasing efficiency on a world scale it is important to take these differences into account. In addition, the emerging economies have massive potential for growth which, if fulfilled, will greatly increase the demand for energy. An assessment therefore must take account of the counter effects of higher efficiencies and increased demand. Many analysts have collated data on energy consumption for various activities in the different economies of the world and overlaid these trends with the impact of policies to curb energy consumption. This chapter takes these projections as a starting point and overlays them with estimates of the impact of achieving the ultimate in energy efficiency from oil and gas.

The scope to improve efficiency in the use of oil and gas is specific to the process, activity or amenity which is being served. Energy saving measures applied to buildings are not often appropriate to vehicles or manufacturing processes. Applications need to be studied in detail to determine how and by how much efficiency can be improved. In this chapter the analysis of potential improvement is conducted under the commonly used headings of industrial usage and commercial buildings; transportation and residential use are not dealt with. For all categories macro data are available for the major economies of the world. These data do not draw out the energy used to generate electricity, a large use of primary fuel. An additional category of potential efficiency improvements, power generation, is therefore included.

8.2 POWER GENERATION

Across the world the electricity generating capacity installed is 2700 GW providing a total annual output of 43 EJ. This represents 12% of energy consumption in the home, industry, commerce and transportation. Central power stations have an average efficiency of 35% in the developed economies and 26% in the developing economies. However, 10% of the electricity generated is dissipated during transmission and distribution. This means the total energy input is some three to four times higher than the electricity consumed. Power generation accounts for 32% of total fossil fuel consumption and 17% of the consumption of oil and gas.

Power stations are generally in the size range 100–2000 MW of electricity (MW_e). Apart from hydro power the fuel input, including nuclear, is generally used to raise steam to drive turbines connected to alternators which generate the electrical output. Recently there has been increased interest in the use of natural gas for power generation and this is accelerating the development of suitable gas turbines. The balance of fuels used for power generation differs geographically and depends on the availability of resources and on the stage of economic development (Figure 8.1). Coal is still the dominant provider of electricity throughout the world and hydro is a major source particularly in the developing economies.

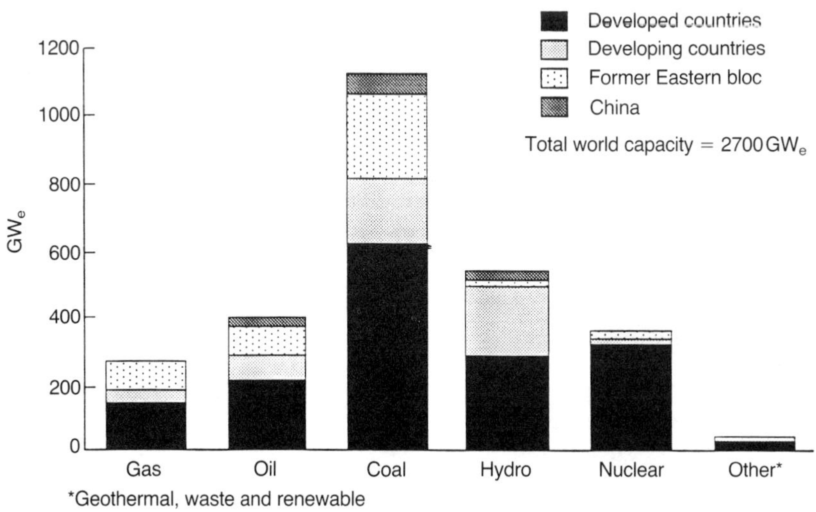

Figure 8.1 Electricity generating capacity by fuel type and economic group.

Nuclear is mainly found in the developed economies. Gas and oil represent 10% and 15% of the total respectively, with natural gas use tending to increase as larger, high efficiency gas-turbine-based systems become available.

World capacity of power generation plant is expected to continue to increase in all economies, but with some uncertainty in the CIS and Eastern Europe. In broad terms an annual growth of 2.6% is predicted for the developed economies to the end of the decade, with 6.6% for the developing economies. This would add about 800 GW_e capacity by the year 2000.

8.2.1 Power generation technology

Prior to 1980, electrical power from fossil fuel was generated almost exclusively from steam-turbine-driven plant. This applied to both central power stations and large industrial sites generating their own heat and power. The steam could be generated using coal, oil or natural gas. With the easing of restrictions on privately generated power, interest grew in the use and development of gas turbines, gas engines, dual fuel engines and fuel cells. These developments are continuing to advance but already there is a wide range of power generation and combined heat and power options available (Table 8.1).

(a) *Steam turbine*

The bulk of power generated at the present time is through Rankine cycle steam turbines with sizes up to 1300 MW. The steam is generated in boilers from either coal, oil, natural gas or nuclear fission. Maximum pressure at the present time is 430 bar with superheat temperatures of 650 °C. This steam is expanded through turbines often with high, medium and low pressure stages. The steam is condensed after the turbine to maximize the pressure differential. The maximum cycle efficiency currently is 45% and this is constrained by metallurgical limits for both the boiler and the turbine on the combination of pressure and temperature which can be used. Advances in materials could increase the efficiency to 47% but this is considered to be the limit for a stand-alone steam cycle plant.

(b) *Gas turbines*

From their inception in the 1930s, stationary gas turbines have been used for peak load power generation. During the past 10 years advances in efficiency, unit size and reliability have enabled gas turbines to be used for base-load power generation, and gas turbines for a broad range of applications are currently available. There are two main families commonly entitled 'industrial' or 'heavy-frame' machines, of up to 230 MW output, and aeroderivative machines of up to 40 MW output (Table 8.2).

Table 8.1 Key parameters of commercially available power generation and combined heat and power plant

Prime mover	*Unit size range*	*Electrical generating efficiency of modern plants (%)*	*Heat to power ratio*	*Installed cost (£/kW)*
Steam turbine	up to 1300 MW_e	20 to 45	0 to 12:1	570 to 685
Small gas turbine	less than 1MW_e	15 to 23	2.5 to 4.5:1	685 to 1140
Large gas turbine	1 to 230 MW_e	20 to 40	1 to 3:1	460 to 860
Combined cycle (gas + steam turbine)	20 to 350 MW_e	35 to 55	0 to 1.5:1	570 to 800
Dual fuel engine	1 to 15 MW_e	33 to 38	1 to 1.5:1	460 to 570
Spark ignition engine	20 kW_e to 2MW_e	33 to 38	1 to 2:1	500 to 800
Phosphoric acid fuel cell	200 kW_e	43	1:1	1700[b]
Not yet commercially available				
Solid polymer fuel cell	5 to 120 kW_e	46	0.8:1	not known[b]
Phosphoric acid fuel cell	50 kW_e to 11 MW_e[a]	43	1:1	1700[b]
Molten carbonate fuel cell	10 kW_e to 2 MW_e[a]	45	0.8:1	1000[b]
Solid oxide fuel cell	20 to 100 kW_e[a]	37	1.5:1	2500[b]

[a] Higher capacity units comprise a number of modules.
[b] Costs for commercial production should be lower.

Table 8.2 Performance parameters of gas turbines

	Industrial gas turbine	*Aeroderivative gas turbine*
Pressure ratio	5–30:1	20–30:1
Turbine inlet temperature (° C)	850–1290	1150–1240
Power output (MW_e)	1–230	0.5–50
Shaft power efficiency (larger units)	20–38	35–40

The principle of operation of both types of turbine is the same. Air is drawn into the compression stage of the turbine. After compression, fuel is introduced and combusted. There is a consequential increase in temperature and volume and the high temperature gases are expanded through the power turbine. The shaft power output is the difference between the power turbine output and the power required for air compression minus any losses.

The industrial gas turbine has evolved from steam turbine technology. It tends to be heavier and larger than the aeroderivative type. The latter is generally more compact and complex, often using two-stage compression and expansion to raise performance. Two shafts are also commonly used to help optimize the compression and power output shaft speeds.

The options available to improve the performance of gas turbines include:

1. improved compressor design;
2. improved power turbine design;
3. higher power turbine inlet temperature;
4. higher pressure ratios;
5. modification of the fluid properties of combustion air.

Significant improvements in the design of the compressor and power turbine can be predicted theoretically but are difficult to engineer. For instance, the compressor absorbs almost two thirds of the power generated by the turbine stage. If effective intercooling could be arranged between compression stages, then compressor power consumption would reduce and net power output could be almost doubled. Arrangements which duct the air through coolers would lead to a lengthening of already highly stressed rotating shafts with the danger of excessive vibration. None the less, such problems should prove tractable even within a decade.

Higher turbine inlet temperatures are strongly linked to higher pressure ratios. It is advances in these parameters which will dramatically improve turbine performance. At the present time the temperature limit is 1290 °C. At higher temperatures, current materials and designs cannot withstand the high stresses which occur within gas turbines.

Figure 8.2 shows how the efficiency increases with turbine inlet temperature and how the optimum pressure ratio is linked to inlet temperature. If suitable materials could be developed and used, possibly in conjunction with improved blade cooling and blade coatings, then an efficiency increase from the current 37% to 45% could be achieved. A step change to ceramic materials may occur within as little as two decades.

An essential requirement of any power plant of the future will be clean exhaust gases. Turbines in the past have had NO_x levels of the order of 150 p.p.m. Current developments are achieving 10 p.p.m. If turbine inlet temperatures rise then so will NO_x levels and above 1530 °C, NO_x levels

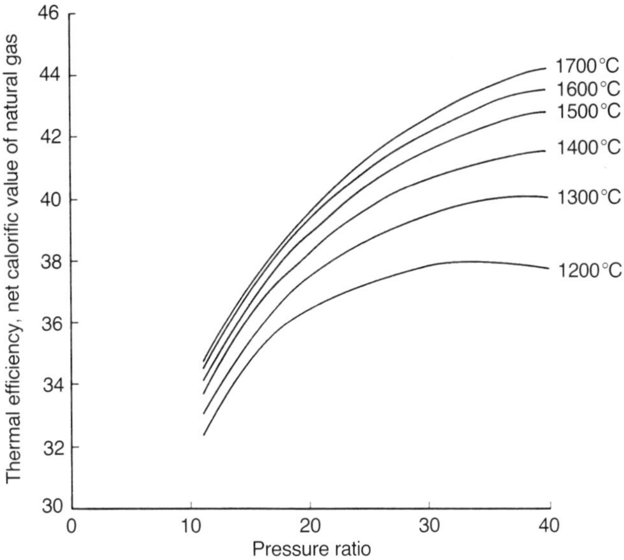

Figure 8.2 Gas turbine thermal efficiency vs. pressure ratio at different turbine inlet temperatures.

rise rapidly. This may be alleviated by catalytic combustion and post-turbine clean-up technology.

Should materials availability, engineering difficulties and NO_x levels associated with higher temperatures prove to be insuperable problems over the third millennium we should not be disheartened. Gas turbine performance can be enhanced in a number of ways and the gas turbine can be configured into a variety of combined cycle systems to make dramatic improvements in performance.

8.2.2 Advanced power generation systems

(a) *Combined cycle gas turbines*

Combined cycle gas turbine systems are the basis of the new generation of natural gas fuelled power stations. The term describes the integration of a gas turbine with a steam turbine such that the hot gases leaving the gas turbine are subsequently used to raise steam for the steam turbine (Figure 8.3). This is possible because the gas turbine converts typically only 30–40% of the fuel input to shaft power. The remaining energy leaves in the form of hot gases at up to 600 °C. These gases can be used in a boiler to raise steam which can then drive a steam turbine to generate additional power.

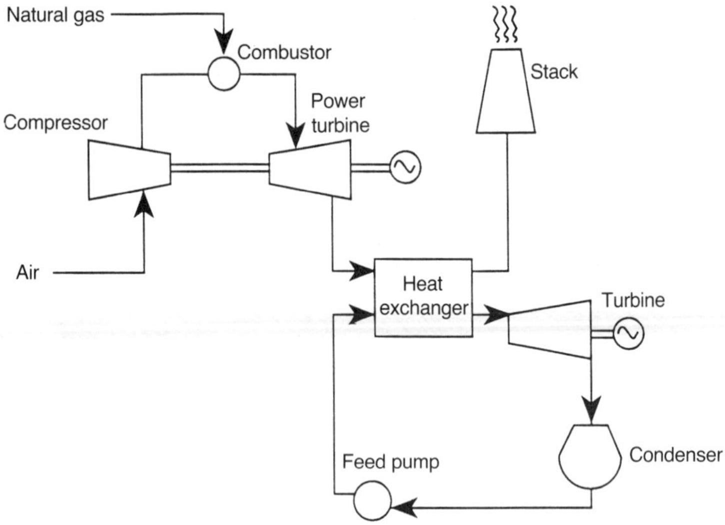

Figure 8.3 Combined cycle gas turbine.

To date, combined cycle gas turbine systems have been built using a variety of combinations of gas and steam turbines of up to 2800 MW. In general there has been no mechanical connection between the gas and steam turbine. Current state-of-the-art technology is, however, creating changes and a packaged unit is becoming commercially available in which the gas turbine, steam turbine and alternator are on a single shaft. This arrangement reduces floor space and foundation costs but gives some reduction in operational flexibility. The power output is up to 340 MW with an electrical efficiency of up to 57%.

(b) *Fuel cells*

Fuel cells are electrochemical devices which can convert the chemical energy of fuel directly into electricity (Figure 8.4). The principle has been known for 150 years but to date the technology has not been generally competitive with the established means of electricity generation. At present large investments are being made to develop the technology in the United States and Japan, with European programmes now starting to become significant.

Fuel cells are defined by their electrolyte and there are four main types being developed for power generation (Table 8.3). These are:

1. phosphoric acid fuel cell;
2. molten carbonate fuel cell;
3. solid oxide fuel cell;
4. solid polymer fuel cell.

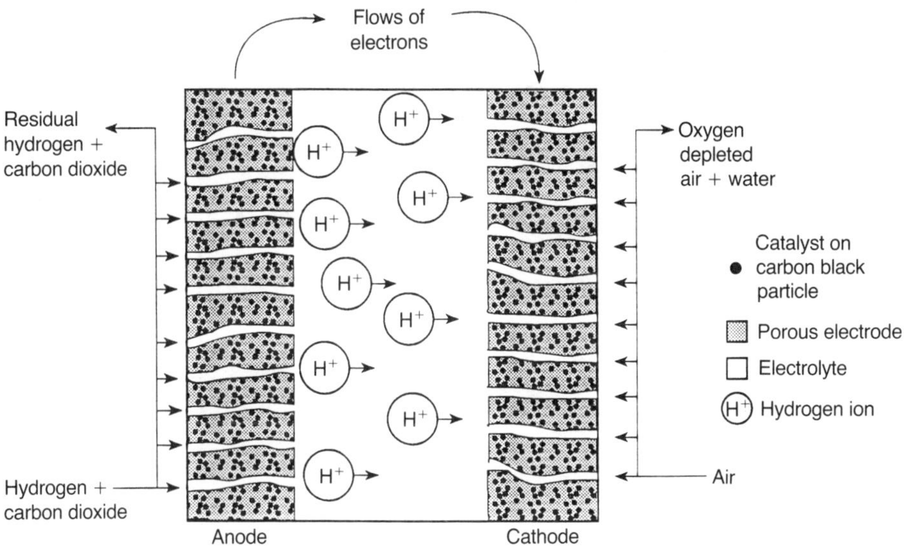

Figure 8.4 Fuel cell (phosphoric acid).

The temperatures of operation for each type differ significantly with the higher temperature molten carbonate and solid oxide units providing scope for incorporation into combined cycle plant. Electrical efficiencies range from 30% to potentially 65%, which makes them an attractive option either as central station plant or as combined heat and power units.

Fuel cell technology is becoming available but is still at an early stage of development. Phosphoric acid 200 kW combined heat and power units are available at £1700/kW and some are being evaluated around the world. Tokyo Electric Power Company is operating an 11 MW phosphoric acid fuel cell power plant consisting of multiple modules. Experimental molten carbonate units of up to 100 kW and solid oxide units of up to 25 kW are in operation. Prices are expected to reduce as the technology becomes established and, if cost targets can be met, fuel cells may become a competitor to, or complementary to, gas turbines in advanced configurations of combined cycle plant.

(c) *Analysis of the potential of combined cycle options to increase thermal efficiency*

So far this section of the chapter has briefly described the type of equipment available to generate electrical power and provided some data on current and potential performance. At this time the turbine-based systems and fuel cells appear to be promising contenders for fossil fuel fired power station

Table 8.3 Features and performances of fuel cell types

Fuel cell type	*Materials used*			*Acceptable fuels*	*Operating temp (°C)*	*Electrical efficiency*	
	Anode	*Cathode*	*Electrolyte*			*Present*	*Limit*
Solid oxide 2 MW_e plant due for 1995	Nickel	Sr doped $LaMnO_3$	Yttria doped zirconia	Hydrogen Natural gas Carbon monoxide	1000	37	60–65
Molten carbonate 2 MW_e units due for 1997	Nickel	Nickel oxide	Molten Li/Na/K carbonate	Hydrogen Natural gas Carbon monoxide	650	45	60–65
Phosphoric acid 11 MW_e unit in operation	Platinum	Platinum	Phosphoric acid	Hydrogen	200	43	50–55
Solid polymer 5 kW_e units available	Carbon supported platinum	Carbon supported platinum	Hydrated organic polymer	Hydrogen	80	46	50–55

plant of the future. The limits of efficiency will therefore be explored using these technologies.

Computer programs have been used to predict the performance of a number of enhanced combined cycle systems based on available turbines but incorporating realistic expectations of future turbine performance. One program contains all the operating data for currently available turbines and allows variations in performance to be assessed for changes in operating conditions. This is used in conjunction with an energy system analysis program which can evaluate different options of combined cycle plant. The more promising options being explored include:

1. steam injected gas turbine;
2. humid air turbine;
3. chemically recuperated turbine;
4. partial oxidation turbine;
5. gas turbine and Kalina cycle;
6. gas turbine and fuel cell.

The schematic arrangements, features and performances for the six options are shown in Figures 8.5, 8.6 and 8.7. Other options and variations are possible but current indications are that realistic extrapolations of the performance of currently known technology should enable power generation efficiencies approaching 80% to be achieved. This is double that presently attained in power generation plant and the rate of progress towards that goal will depend upon economic and environmental factors.

8.2.3 Combined heat and power (CHP)

An alternative to central power generation is combined heat and power. This is the generation of power on an industrial, commercial or residential site with the heat recovered from the prime mover being used for process, space or water heating. With current CHP technology the efficiency of electricity generation is about 30% with up to 60% of the input energy recovered as heat. This gives overall efficiencies of over 80%. The technology is finding favour around the world and there are 9 GW of gas fuelled plant installed in the United Kingdom, which represents 1.2% of UK power generation capacity.

Combined heat and power offers the prospect of distributed power generation from a great number of small units linked to a national grid. Such an arrangement would have implications for the electricity transmission system in that its main purpose would become one of balancing local supply and demand mismatches, rather than transmitting power over long distances. In some distant future one could envisage a mix of power generation facilities including combined heat and power, wind, wave, hydro and solar.

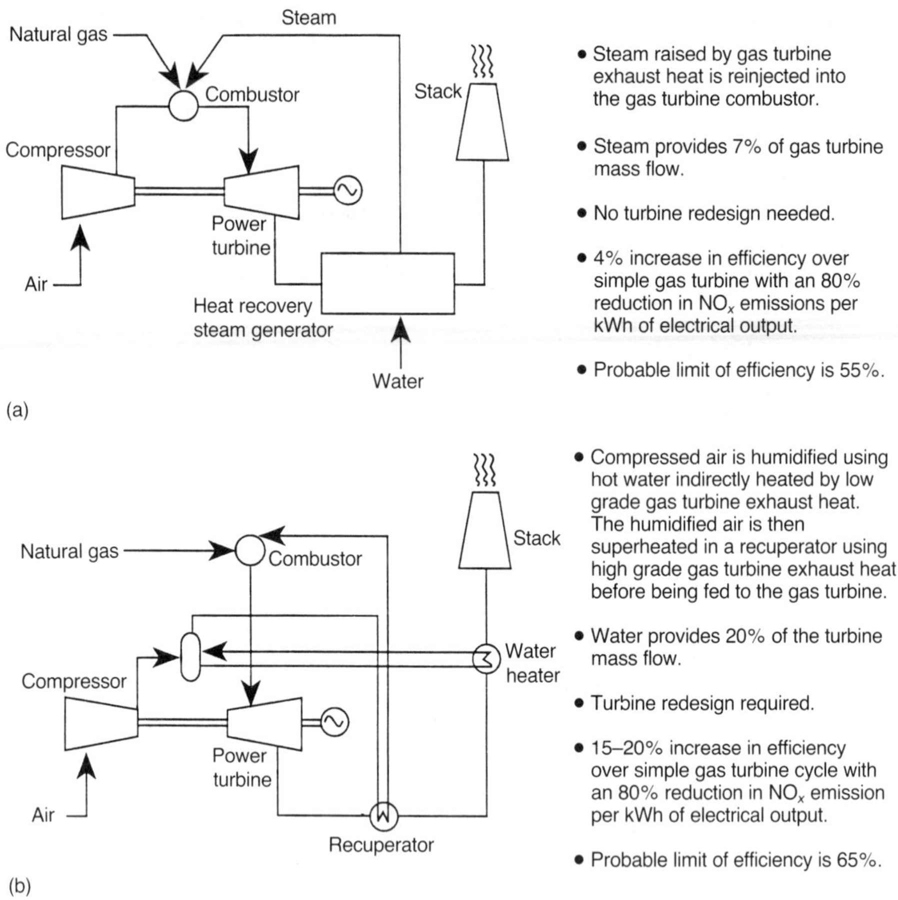

Figure 8.5 (a) Steam injected turbine; (b) humidified air turbine.

The average efficiency of electricity generation in central power stations in the United Kingdom in 1992 was 32% after transmission. Comparing energy use in a combined heat and power system with the conventional model of buying in electricity and burning fuel in a boiler, the ratio is 1 to 1.5. That is, the combined heat and power system saves about 34% of fuel. The efficiency of fuel-fired central power stations would need to rise to about 65% to achieve parity on fuel use with existing combined heat and power equipment. Any technological advances which brought that about would also be reflected in CHP technology. There would therefore be a theoretical convergence somewhere above 80%.

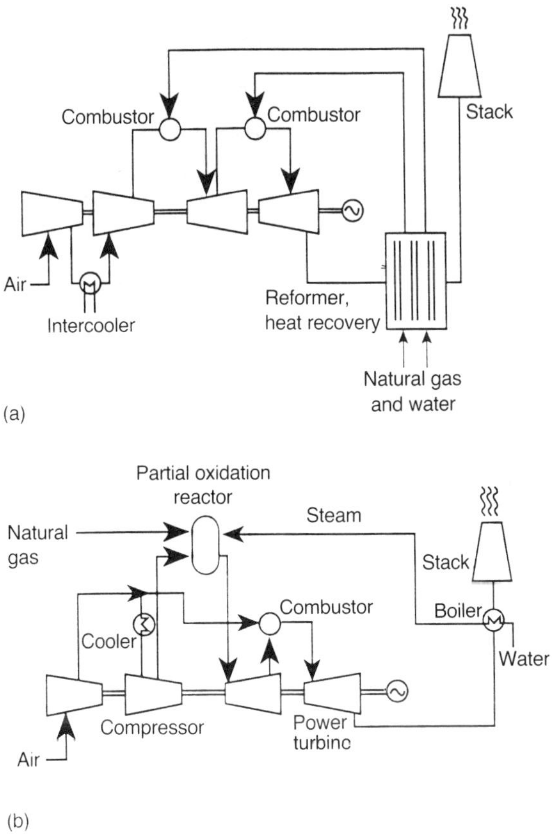

- The natural gas feed is mixed, with water, preheated and reformed to a hydrogen/carbon monoxide/carbon dioxide/methane/steam mixture which is used hot as fuel for the gas turbine combustor. The heat required for pre-heating and reforming is provided by the gas turbine exhaust.
- Engineering development needed.
- 15–20% increase in efficiency over simple gas turbine cycle with an 80% reduction in NO_x emissions per kWh of electrical output.
- Probable limit of efficiency is 65%.

- The natural gas is partially oxidised in a reactor vessel to a carbon monoxide/hydrogen rich mixture. The reaction products are then expanded through an unfired power turbine before complete combustion with the remaining air, and then expansion in a power turbine.
- Engineering development needed.
- 15–20% increase in efficiency over simple cycle gas turbine with an 80% reduction in NO_x emissions per kWh of electrical output.
- Probable limit of efficiency is 65%.

Figure 8.6 (a) Chemically recuperated turbine; (b) partial oxidation turbine.

8.2.4 The impact on world energy demand of achieving the realizable limits of efficiency for power generation

In the developed economies the average efficiency of power generation is 35%; it is 26% in the developing economies. The energy sources are a mix of coal, oil, gas, nuclear, hydro and renewables. The nature of this mix will change with time, depending on economic, technological, environmental and political factors. In practice, therefore, projections of the future can only describe a boundary to the possibilities should gas and oil technology for power generation achieve the realizable limits. A number of options have been evaluated for varying scenarios of growth and equipment replacement (Figures 8.8(a),(b) and (c)).

Figure 8.8(a) compares current power generating capacity with the

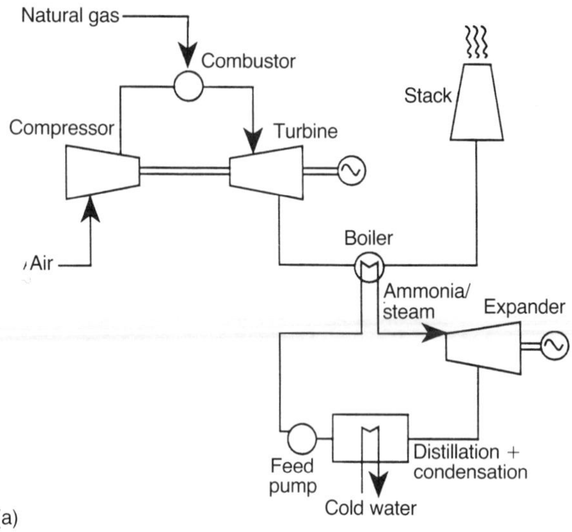

(a)

- Gas turbine exhaust heat boils ammonia/water mixture in a condensing steam turbine type cycle.
- No turbine redesign needed.
- 20% increase in efficiency over simple gas turbine cycle with a 20% reduction in NO_x per kWh of electrical output.
- Probable limit of efficiency is 62%.

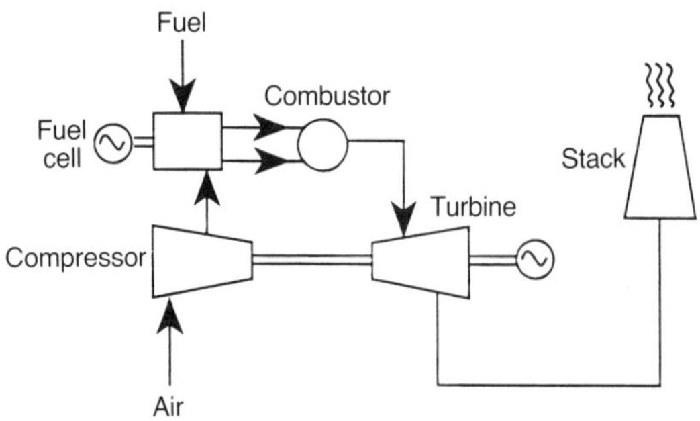

(b)

- 85–90% of the fuel is reacted in a high temperature, high pressure fuel cell. The unreacted fuel is combusted with the hot fuel cell cathode exhaust and fed to a gas turbine. The turbine exhaust heat drives a steam cycle.
- Turbine combustor modifications needed. Fuel cell scale-up and pressurized design needed.
- 40% increase in efficiency over simple gas turbine cycle with a 90% reduction in NO_x emissions per kWh of electrical output.
- Probable limit of efficiency is 75%.

Figure 8.7 (a) Gas turbine with Kalina cycle; (b) gas turbine with fuel cell.

expected level by the year 2000, assuming World Bank growth rates and that the current fuel mix and efficiencies are maintained. Fossil fuel and carbon dioxide emissions rise by 29%.

Figure 8.8(b) assumes that power generating capacity remains constant. The upper bar shows current fuel mix and efficiency whilst the lower bar gives the effect of replacing all non-renewable capacity with 80% efficient oil/gas plant. Fossil fuel consumption would be almost halved and CO_2 emissions cut to one third. The consumption of oil and gas would rise by 10%.

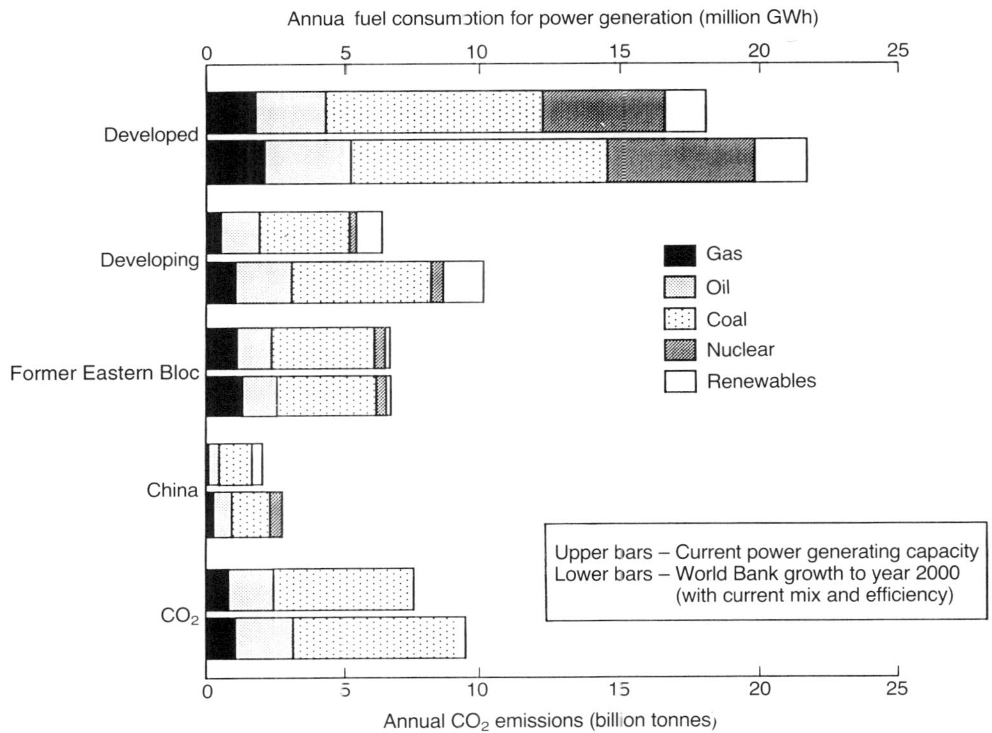

Figure 8.8 (a) Impact on power generating capacity of growth to year 2000 with current mix and efficiency. (b) Impact on current power generating capacity of realizable efficiencies. (c) Impact on power generating capacity of all economies reaching OECD 2000 levels.

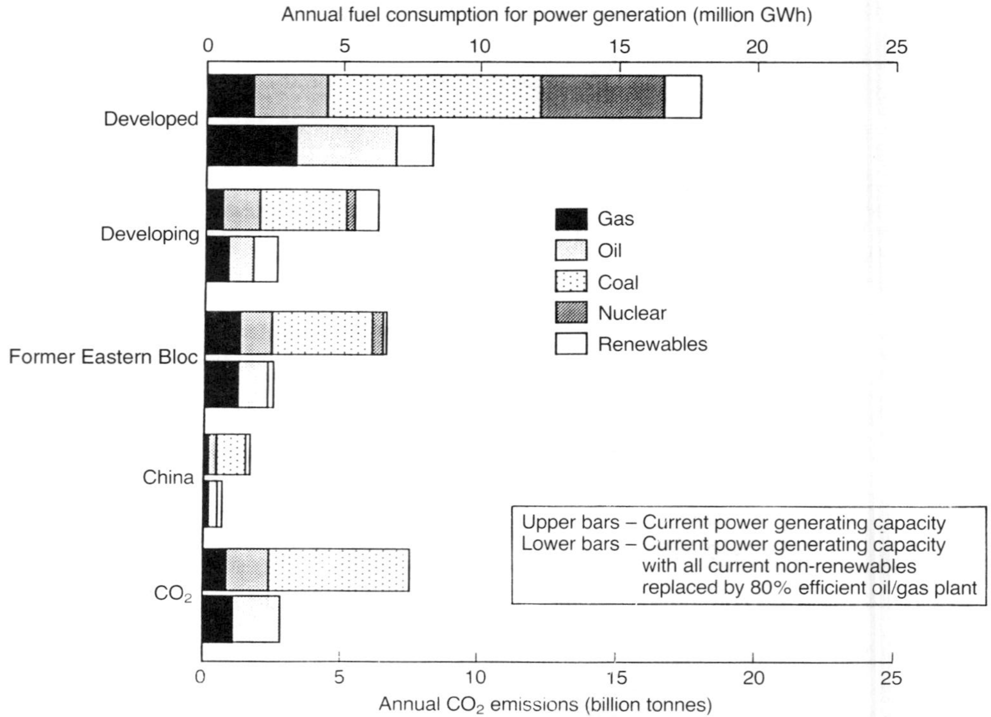

Annual fuel consumption for power generation (million GWh)
0
5
10
15
20
25
Developed
Developing
Former Eastern Bloc
China
CO_2
Gas
Oil
Coal
Nuclear
Renewables
Upper bars – Current power generating capacity
Lower bars – Current power generating capacity with all current non-renewables replaced by 80% efficient oil/gas plant
Annual CO_2 emissions (billion tonnes)

Figure 8.8 (b)

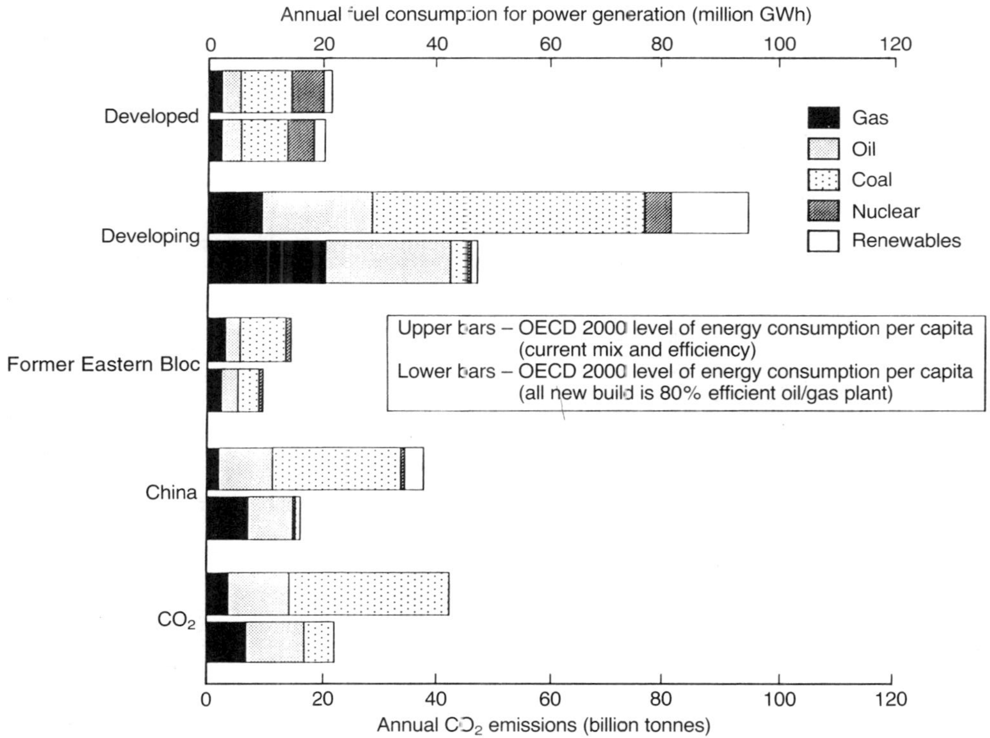
Annual fuel consumption for power generation (million GWh)
0
20
40
60
80
100
120
Developed
Developing
Former Eastern Bloc
China
CO2
Gas
Oil
Coal
Nuclear
Renewables
Upper bars – OECD 2000 level of energy consumption per capita (current mix and efficiency)
Lower bars – OECD 2000 level of energy consumption per capita (all new build is 80% efficient oil/gas plant)
Annual CO2 emissions (billion tonnes)

Figure 8.8 (c)

Figure 8.8(c) compares two long-term scenarios of all economies achieving OECD levels of electricity consumption per capita. The upper bars are based on the assumption that the growth occurs with the current fuel mix and efficiency. If this were the case, fossil fuel consumption would rise by 300% and CO_2 emissions by 480%. The lower bar assumes the growth is provided by the addition of 80% efficient oil- and gas-fired plant. If this were the case, fossil fuel consumption would rise by 180% and CO_2 emissions by 190%. The consumption of oil and gas would rise by 510% for power generation.

8.2.5 Conclusions – power generation

1. Power generation technology based solely on the steam cycle is unlikely to achieve electrical efficiencies of 50%.
2. Power generation technology based on gas turbine systems is now becoming available with electrical efficiencies of 58%.
3. Evolutionary development of technologies such as gas turbines and fuel cells, configured into advanced combined cycle systems and combined heat and power systems, should enable limiting efficiencies of about 80% to be achieved during the next century.
4. Growth in primary energy use for power generation could be halved by the use of generating capacity operating at the limits of efficiency.
5. Electricity demand is expected to grow world-wide. The impact of this growth will outweigh the potential energy savings such that energy use will rise.

8.3 INDUSTRIAL ENERGY USE

The ratio of manufacturing to total energy use is different for the various economies of the world. It is 40% for the world, 30% for the OECD countries, 40% for the former USSR, 60% for China and 35–45% for the emerging economies. These ratios change with time owing to changing consumptions in other sectors such as transport, housing and commerce, and due to the restructuring, relocation and growth of the global industrial base.

The energy used for manufacturing has changed with time, and data are presented for 1970 to 1988 (Figure 8.9). The consumption in the OECD countries has fluctuated significantly. It is difficult to isolate the reasons for this but they can be implied. Oil shortages and price variations will have had some impact, but so will variations in output, industrial restructuring and rationalization, as well as the use of more productive and efficient plant and materials substitution. The net result is that energy consumption in the OECD countries was lower in 1988 than in 1973 despite an increase in the value of manufactured goods of over 30%.

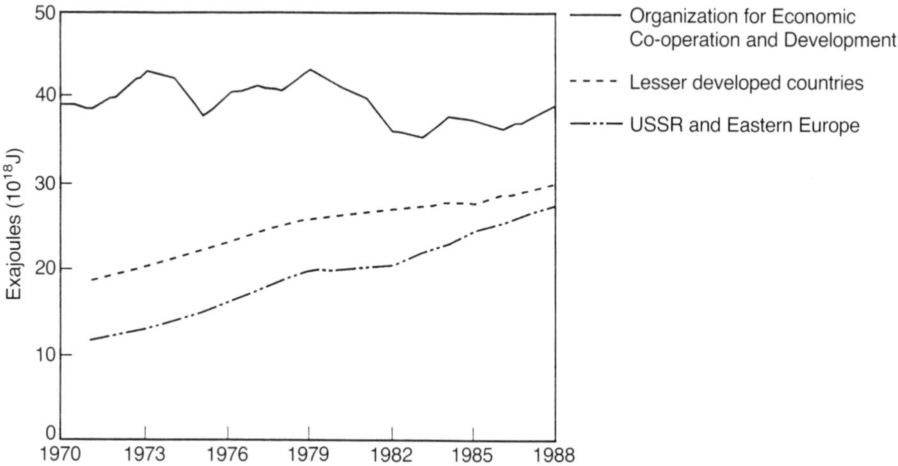

Figure 8.9 Annual industrial energy use 1970–1988 – includes mining, construction and feedstocks. Source: IEA, after Schipper and Meyer (1992).

The former USSR and the emerging economies have shown a steady increase in industrial energy consumption. In the former USSR there has been a concentration on increasing production from heavy industries without any major improvements in production methods. The steady increase in energy consumption therefore matches growth in production. The emerging economies differ widely in industrial growth rates, and in their use of old or modern manufacturing plant. Also, some countries have placed emphasis on energy saving. The steady upward trend in energy consumption therefore reflects a general increase in industrial output but masks a wide variety of practices and growth rates in industrial activity.

8.3.1 Energy use by industrial sector in the OECD-8 countries

Of total world industrial energy consumption, 70% occurs in five energy-intensive industries: ferrous metals; non-ferrous metals; chemicals; paper and pulp; stone, clay and glass. All five industries are large energy users but collectively they only add 20% of added value. Other industries include mining, construction, food, textiles, ceramics, rubber, plastics and engineering, which produce high value products with relatively low energy consumptions.

The efficiency of energy use can to some extent be indicated by the energy consumption per tonne. The five broad categories of industries, however, include a wide range of types of products which require different energy inputs during processing. An example is construction steel, which requires a different amount of heating and rolling compared to thin sheet for food

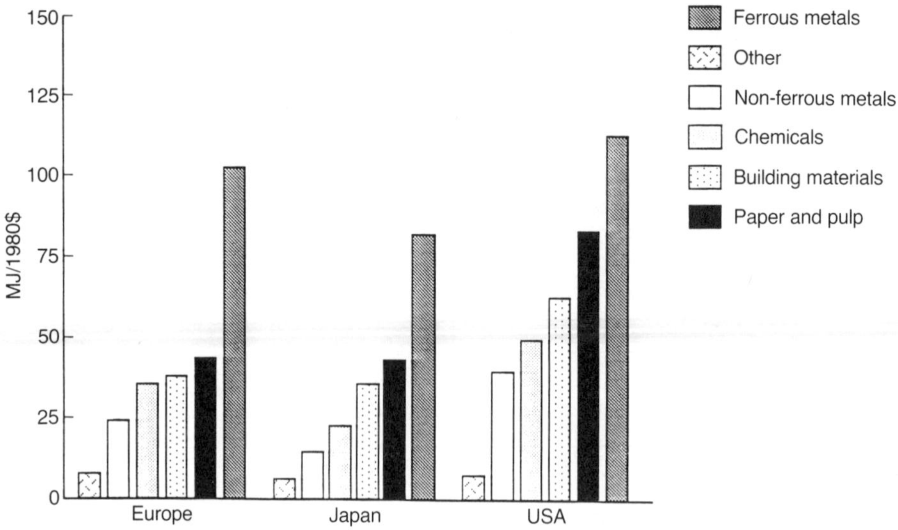

Figure 8.10 Energy intensity by industrial group for the United States, Europe-6 and Japan (1988).

cans. A more common approach is to define energy intensity as energy per unit of value added. Thus energy intensity is defined as MJ/1980$. Again, this is not a precise basis for comparing production practices, but uses the best data we have available.

In a comparison of energy intensity for the five sectors for the United States, Europe-6 and Japan, the United States has the highest intensities and Japan the lowest (Figure 8.10). These data may indicate that energy is used more efficiently in Japan to produce a given product. They may also indicate that Japan is producing a higher proportion of high value added products. Both reasons are probably valid reflecting the high cost of energy in Japan as well as the drive to move continuously into new, high value products.

8.3.2 Energy use by the industrial sector in the United Kingdom

The UK industrial energy market totalled 1.5 EJ in 1990 and accounted for 27.3% of UK energy consumption. This is a reduction of 34% from 1980. The energy usage is categorized as low temperature heating processes (<200 °C), high temperature processes (>200 °C) and industrial buildings (Figure 8.11). The temperature of 200 °C is a convenient boundary for process technologies and for some industries.

A survey was carried out in 1990 which quantified the flow of gas, oil, solid fuel and electricity into eight industrial groupings (Table 8.4). It also

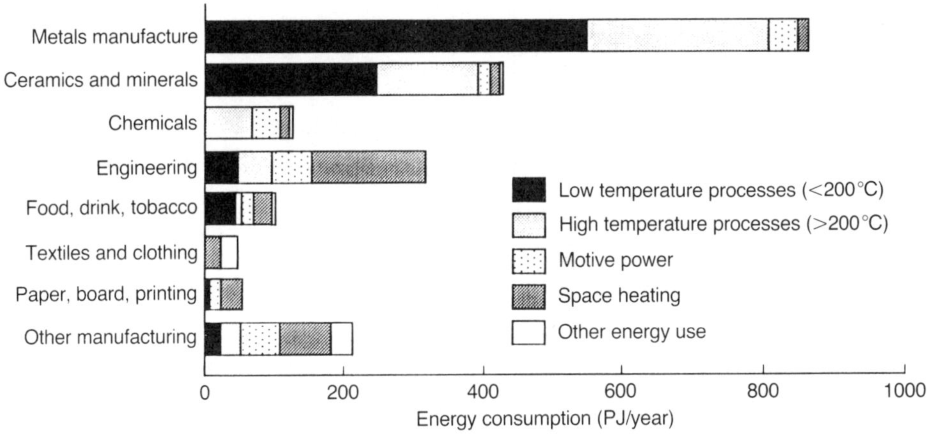

Figure 8.11 Distribution of energy use within UK industry.

provided data in broad terms on the type of application in which each fuel is used. These data are the starting point against which the effects of improvements in energy technology on UK industrial fuel consumption are based.

8.3.3 Low temperature process heating (<200 °C)

A total of 540 PJ of energy were used in this sector in 1990. Process heating by centrally generated steam from boilers and steam-driven CHP accounted for 80% of this, gas-fired CHP for 0.6%, direct firing of processes 11.5%, and the balance of 7.9% was for stationary motive power, mainly electricity, for pumps, fans and production plant.

The four basic processes in this sector are liquid heating, drying, evaporation and distillation. Liquid heating refers predominantly to aqueous solutions and these are used for washing, coating, pickling and cleaning processes. Heating is generally by a steam coil immersed in the liquid or by direct injection of steam. In recent years high efficiency direct-fired gas systems have been developed for steam replacement.

Drying is the removal of water or other volatile liquids from solids or powders by evaporation. It is an energy-intensive process and has been the subject of much research. Generally, thermal drying is more costly in energy than mechanical means and the selection of drying equipment should always take into account the scope for upstream dewatering processes. A good example of the trend towards mechanical dewatering is the paper industry. Typically the final drying stages remove 3% of the water with 75% of the energy. Significant advances are being made in pressing

Table 8.4 Fuel use in UK manufacturing industry

	1990 Total use (GWh/year)	*Percentage of total energy use*	*1990 Total use (GWh/year)*	*Percentage of total energy use*	*Main users*
Industry sector					
Consumption of natural gas					
Metals extraction and manufacture	19 313	22.3	18 523	19.4	Steam use in blast furnaces, secondary metal finishing, post-finishing operations, direct fired equipment
Ceramics and mineral industries	13 828	23.2	18 874	40.1	Direct fired equipment, space heating, steam raising
Chemicals	15 269	27.8	13 894	27.3	Steam raising, direct fired equipment, space heating
Engineering	30 919	31.5	27 314	46.4	Direct fired equipment, steam raising, space heating
Food, drink and tobacco	15 152	29.0	19 167	42.9	Steam raising and space heating, direct firing (e.g. baking)
Textiles and clothing	5 686	22.0	5 744	43.7	Steam raising, space heating
Paper, board and printing	8 675	26.8	9 847	38.6	Steam raising space heating
Other manufacturing industries	11 781	19.6	10 961	18.9	Steam raising, space heating
TOTAL INDUSTRIAL GAS USE	120 423	25.7	124 324	31.6	
Consumption of liquid fuels					
Metals extraction and manufacture	20 602	23.7	9 829	10.3	Steam use in blast furnaces, secondary metal finishing, post-finishing operations, direct fired equipment
Ceramics and minerals industries	17 160	29.2	9 536	20.3	Direct fired equipment, space heating, steam raising
Chemicals	18 307	33.3	11 843	233	Steam raising, direct fired equipment space heating
Engineering	26 262	26.8	9 587	16.3	Direct fired equipment, steam raising, space heating, motive power
Food, drink and tobacco	24 018	46.0	11 334	25.4	Steam raising and space heating
Textiles and clothing	12 239	47.3	2 958	22.5	Steam raising, space heating
Paper, board and printing	12 251	37.9	3 111	12.2	Steam raising, space heating

Other manufacturing industries	24 503	40.9	29 946	51.5	Steam raising, direct fired equipment, motive power, space heating, off-road vehicles
TOTAL INDUSTRIAL OIL USE	155 342	33.1	88 144	22.4	
Consumption of coal and solid fuels					
Metals extraction and manufacture	36 781	42.4	51 229	53.7	Blast furnace, production of iron castings, direct fired equipment
Ceramics and minerals industries	22 596	38.5	11 225	23.9	Direct fired equipment
Chemicals	3 546	6.4	7 766	15.3	Steam raising
Engineering	15 123	15.4	2 081	3.5	Space heating
Food, drink and tobacco	6 242	11.9	3 370	7.5	Steam raising and space heating
Textiles and clothing	3 546	13.7	1 729	13.2	–
Paper, board and printing	5 715	17.7	4 572	17.9	Steam raising
Other Manufacturing Industries	4 396	7.3	1 172	2.0	Direct fired equipment, steam raising, space heating
TOTAL INDUSTRIAL SOLID FUEL USE	97 946	20.9	83 145	21.1	
Consumption of electricity					
Metals extraction and manufacture	10 061	11.6	15 810	16.6	Electric arc steel making, post-finishing operations, aluminium production
Ceramics and minerals industries	5 361	9.1	7 423	15.8	Motive power, lighting, melting processes
Chemicals	17 889	32.5	17 351	34.1	Motive power, electrochemical processes, lighting
Engineering	25 780	26.3	19 925	33.8	Motive power, direct process use, lighting
Food, drink and tobacco	6 837	3.1	10 834	24.2	Motive power, refrigeration, lighting
Textiles and clothing	4 416	17.1	2 700	20.6	Motive power
Paper, board and printing	5 682	17.6	8 011	31.4	Motive power
Other manufacturing industries	19 293	32.2	16 053	27.6	Motive power, lighting, space heating
TOTAL INDUSTRIAL ELECTRICAL USE	95 319	20.3	98 107	24.9	

Sources: Eurostat (All fuel data excluding note below). *Digest of UK Energy Statistics* (electricity, solid fuels for iron and steel and 1980 data).

technology, with slower progress on novel drying techniques and dry paper-making techniques.

Evaporation is the removal of volatile liquids from a solution of other less volatile liquids which may contain dissolved solids. Evaporation systems often have high performance with the vapours from one stage being used to heat a subsequent stage by condensation. Modern systems can have up to 10 stages, and significant advances have been made over the past 20 years in the design of heat transfer surfaces and the reduction in temperature driving forces. The energy requirement for a typical evaporation process has reduced by 30–50% over the past 20 years. Some advances will continue to be made but significant further improvements will probably come from advances in the technology of heat pumps.

Distillation is the separation of liquid mixtures into several distinct components with high levels of purity. The separation is achieved by selectively heating and cooling parts of a tall vertical distillation column, most notably the products at the top and bottom stages. Good design is crucial to high thermal performance and improved models and process integration techniques have enabled major advances to be made. Future advances should come from dynamic modelling and improved control systems.

The process plant which use the above phenomena are often complex with many process operations being carried out. They usually involve the heating or cooling of liquid or gaseous streams, or vessels in contact with solids. Heat exchange or process transformations can take place in a variety of sealed heat exchangers, pressurized retorts, cyclones and open vessels. There is often scope for heat exchange within the process as well as waste heat sources which can be utilized. Finally there is a significant demand for electrical power. These processes therefore lend themselves to process integration techniques, process intensification, combined heat and power, heat pumps and steam displacement by direct firing.

(a) *Energy saving technologies for low temperature heating processes*

Process integration

Process integration is the optimal configuration of an industrial process such that the process streams, at different stages in the process, exchange heat with each other so as to provide heating and cooling rather than using external heating and cooling media. The analysis involves complex techniques using computer programs to optimize process configurations, temperature differentials, energy flows and the impact on capital costs.

Some 24 process integration studies have been published for the United Kingdom (Table 8.5), and British Gas has conducted a further seven (Table 8.6). Savings range from 13% to 45% using current technology. The scope

Table 8.5 Published UK process integration studies

	Number of studies	*Energy saving (%)*	*Energy saving £million/year)*
Chemicals	10	34	15.6
Oil refining	2	26.6	6.8
Food and drink	7	28.8	4.3
Textiles	1	19	0.2
Paper and board	2	22.6	0.6
Iron and steel	2	13.5	2.9

Source: Mercer A.C., 'Experiences of the Application of Process Integration in UK industry'. Paper presented to the IEA Workshop on Process Integration, Gothenburg, Sweden 1992.

Table 8.6 British Gas process integration studies

	Energy saving (%)	*Energy saving (£k/year)*	*Achievable future savings (%)*
Brewery	44.3	1400	77.5
Whisky distillery	29.5	379	68.0
Chemical works 1	25.3	403	Not known
Chemical works 2	13.6	2671	30.0
Chemical works 3	16.7	883	Not known
Food processor	14.0	62	32.9
Paper mill	25.0	600	40.0

to make savings is site specific and the simple payback periods range from a few months to 5 years. The value of the energy savings can be very high and as plant is replaced over the years, these levels of savings should become the norm.

Combined heat and power

The technology of combined heat and power is well established as discussed in section 8.2. In fact, The Netherlands produces 40% of industrial electricity demand by CHP. A prime target for its use are industries with a high power requirement and a need for low temperature process heating. Good progress has been made during the past decade in the United Kingdom and currently 89 schemes are installed in a variety of sizes and applications. Total capacity is 1619 MW, with 87% being steam turbines, 13% being gas turbines. With current economics and technology the market in UK industry is estimated to be 5000 MW. Advanced technology could increase this to 8000 MW as lower heat to power ratios would provide a suitable match in a wider range of applications.

Heat pumps

Fossil fuel engine-driven heat pumps, primarily driven by gas engines, have been assessed over the past decade in a number of process uses. Many were technically successful giving ratios of heat output to fuel input of over 2:1. Capital costs were high, fuel prices fell and the technology went into an abeyance. Over a long timescale the technology should re-emerge and could offer much higher performance driven by advanced high efficiency prime movers and fuel cells.

Process intensification

Process intensification is the technique by which thermal and chemical processes and operations can be intensified, combined and integrated to produce novel, more efficient and substantially smaller equipment. The primary benefits are reductions in weight, inventory, residence times and cost. There may also be reductions in energy consumption. Examples of process intensification include compact heat exchangers, compact reformers, the hydro cyclone which utilizes a rotation field to enhance liquid–liquid separation, and the Higee distillation unit which enhances gas–liquid mass transfer, again by utilizing a rotational field.

(b) *The impact on energy demand on achieving the realizable limits of efficiency for low temperature heating processes*

An idealized study would take all products currently produced, evaluate the minimum energy required for each and calculate the total minimum energy demand. The pattern of use is continually changing, however, and the data to carry out such a study are not available. An alternative approach is to gather data on existing energy use, determine the typical energy consumption related to current good practice, carry out some representative studies which show the scope for improvement using new available technology and overlay the results with the impact of advanced technology. This is the type of analysis which has been attempted.

Table 8.5 shows the energy savings for each industry sector where process integration studies have been published. The energy cost savings were achieved by savings in fuel use, for example by increased heat recovery, reductions in steam consumption and reductions in electrical use. Some cost savings were achieved from CHP, mostly by the reduction in the use of purchased electricity.

For the processes studied by British Gas the total energy savings range from 14.0% to 44.3% (Table 8.6). The techniques used to reduce energy consumption can be applied across all the low temperature heating industries although the level of savings will be process and site specific.

Combining data from British Gas (Table 8.6) with those from published process integration studies (Table 8.5) provides a tentative basis for an estimate of the UK potential to save energy. In essence the assumptions are:

1. The advanced process technologies can be applied generally across these process industries.
2. The advanced process technologies will yield similar levels of savings in most applications.

Table 8.7 summarizes the potential for energy savings by traditional energy management and equipment replacement, process integration and novel technologies. The first column shows industrial energy use in 1980. The second uses Energy Efficiency Office predictions for energy use in the year 2000. This value is based on a low production increase and poor investment scenarios throughout the 1980s and 1990s. The change of energy use predicted by this scenario is compared with 1980 consumption in the third column.

The next two columns identify the scope for process integration as an energy saving measure. The contribution of process integration has been estimated using published data on UK studies (Table 8.5) and information from British Gas customers. The first of these columns estimates the numerical value for the potential impact of process integration while the second column shows the importance of this contribution to industrial energy consumption.

The next two columns of Table 8.7 attempt to extrapolate existing savings to those felt likely to be achieved in the longer term. The basis for this extrapolation is the result of the British Gas process integration studies (Table 8.6). The total savings from all three measures are shown in the final column.

The results show potential energy savings from 1980 to 2000 to total 115 900 GWh/year or 26% assuming a low growth in the UK economy during this period with a corresponding low investment in new and more efficient process equipment. Process integration with current technology incorporating CHP could save an additional 56 000 GWh/year bringing total savings to 38%. Many of the savings using process integration will result from identifying the most cost effective of the current energy efficiency technologies. The use of process integration and advanced technology could save a further 49 110 GWh/year energy demand, bringing total savings to 50%.

Table 8.7 Scope for energy efficiency from process integration and advanced technology within UK industry[a]

Industry sector	*Energy use in UK industry 1980 (GWh/year)*	*Estimated energy savings by 2000 (low scenario) (GWh/year)*	*Percentage energy savings (from 1980) (%)*	*Potential for process integration (1980–2000) (GWh/year)*	*Energy savings from process integration (1980–2000) (%)*	*Long-term savings from advanced technologies (GWh/year)*	*Energy savings from advanced technology (%)*	*Total potential energy savings (%)*
Metal manufacturing	92 800	27 800	30.0	9 000	9.7	4 500	4.8	44.5
Ceramics	59 500	17 600	29.6	5 600	9.4	2 800	4.7	43.7
Chemicals	90 000	27 200	30.2	25 900	28.8	25 900	28.8	87.8
Paper and board	37 600	8 800	23.4	4 600	12.2	2 760	7.4	43.0
Food and drink	48 800	10 700	21.9	6 400	13.1	6 400	13.1	48.1
Engineering	84 400	17 000	20.1	3 100	3.7	4 650	5.5	29.3
Textiles	33 400	6 800	20.4	1 400	4.2	2 100	6.3	30.9
TOTAL	446 500	115 900	26.0	56 000	12.5	49 110	11.0	49.5

[a] Estimated using data from IEA Seminar, Gothenburg and statistical analysis of UK industry's current energy demands.

8.3.4 High temperature process heating (>200 °C)

A total of 607.2 PJ of energy were used in this sector in 1990. The iron and steel sectors are the largest users of energy. Direct firing of fuel within processes is generally the largest use. Gas accounts for 27.0% of the total energy use, and oil for 15.6%. Coke and breeze remains the highest used fuel at 45.5%. Together these fuels take the largest share, but 11.8% of electricity is also used for process heating which in primary energy terms represents a similar consumption to gas and oil.

High temperature processes are extremely diverse and can involve heating solids, liquids and powders to temperatures of up to 1600 °C. The size, shape, material, handling needs, furnace structure and temperature of operation are markedly different for the many applications. Furnace designs are product specific and thermal inputs differ by two or three orders of magnitude.

Despite this diversity, fuel-fired furnaces do have common operating principles. A furnace or kiln is a refractory lined chamber into which, or through which, material is placed or transported. The walls, roof and hearth are of a material selected to withstand the hot face temperature but which may also have to tolerate mechanical loading, thermal stress, abrasion, corrosion, chemical attack and penetration by gases or molten materials. Heat is provided by single or multiple burners which have to be located to ensure correct temperature, gas flow and heating patterns. Fuel and air are brought to the burners and combustion takes place either within the burner tunnels or in the furnace chamber. Adiabatic flame temperature using ambient air is above 1900 °C and heat is transmitted from the hot gases by radiation and convection to the walls and product material. The flue gas temperature is therefore higher than the furnace chamber temperature at exit. Furnaces can be from less than 1.0 m^3 up to 1000 m^3 in volume.

(a) *Energy saving technologies for high temperature heating processes*

Options to increase the thermal efficiency of furnaces include:

1. Heat recovery from flue gases into the product stream.
2. Heat recovery from flue gases into combustion air and fuel.
3. Highly insulating lightweight furnace structures to reduce conduction and storage losses.
4. Oxygen enriched combustion air.
5. Advanced control systems to maintain precise air/fuel ratio and temperature control, and minimum thermal input under all conditions.

Heat recovery from flue gases into the product stream

This approach essentially defines a furnace or kiln as a counterflow heat exchanger in which combustion flue gases form the hot stream and the manufacturing product forms the cold stream. It has been commonly practised in steel reheating where steel billets are transported in counterflow to the hot gases by walking beam, pusher or roller mechanisms. In ceramics manufacture the ware is stacked on wheeled bogies which carry the ware through the kiln. In cement manufacture the granules are tumbled through a long inclined, rotating cylinder in counterflow to a stream of hot combustion products.

With this type of heat recovery the thermal capacity of the product stream is sufficient to extract all the heat from the combustion gases. The limitation is usually the practical length of the furnace. This is a function of capital cost, handling complexity, increasing thermal conduction and storage losses together with size and space restrictions on a given site.

Heat recovery from flue gases into combustion air and fuel

This approach can be used on both batch and continuous furnaces and can be used in conjunction with load preheating. In its simplest form, hot flue gases are drawn through or over tubes in a heat exchanger with the combustion air forming the cold stream. Flue gas temperatures can be about 1600 °C and early recuperators were limited to air preheat temperatures of below 500 °C in order to keep within their service limits. Even with this modest level of heat recovery, energy savings could exceed 20%.

Some 20 years ago recuperative burners came into common use. These provided a compact, relatively high performance recuperator, integral with a burner. This gave added flexibility in the location of burners and their associated flueing ports. These devices offered combustion air preheat temperatures of up to 600 °C and fuel savings of up to 40%.

Regenerative systems for heating combustion air have been in use for over 100 years. They have been used predominantly on very large, glass melting furnaces, blast furnaces and on the now defunct open hearth steel melting furnaces. In use, the flue gases pass through a chequerwork of bricks in a sealed chamber for periods of up to 30 minutes. The air flow is then reversed and cold air enters this hot chequerwork and is preheated prior to combustion. The flue gases exit through the previously used and cooled regenerator and raise its temperature prior to the next cycle.

Traditional regenerators are massive structures which are considerably larger than the large furnaces that they serve. Some 10 years ago compact designs which could be retrofitted to the existing population of general furnaces and kilns came into common use. Cycling times were reduced to about 2 minutes, specific size was reduced by two orders of magnitude and air preheat temperatures of 1300 °C could be obtained giving fuel savings of up to 60%. The effectiveness of heat recovery is now up to 95%. No doubt these devices will advance in terms of size, cost and application knowledge but they already demonstrate that the limits of effectiveness for heat recovery into the combustion air can be achieved.

Heat can also be recovered from the flue gases into the fuel stream. However, if heated above about 600 °C, natural gas cracks, causing sooting and blockage of the fuel ports. For natural gas with an ideal air: fuel ratio of about 10:1, a specific gravity (ref. air) of 0.6 and a specific heat of 3250 kJ/(kg K) this represents only one twentieth of the potential of combustion air preheat. For these reasons gas preheating is not widely practised. Oil can be preheated to typically 450–500 °C but again this only represents 6% of the value of combustion air preheating. The heat content in the flue gases will always exceed the thermal capacity of the combustion air. This puts a theoretical limit of 80% on the heat recoverable from flue gases by an air heat exchanger.

Highly insulating lightweight furnace linings

The traditional furnace materials are refractory bricks which have a range of densities, conductivities, temperature limits, strengths and corrosion resistances. Correct selection is essential and specialist materials have been developed to contain materials such as molten steel and aluminium and to provide a high level of refractoriness or insulation. About 20 years ago high temperature furnace linings made from lightweight ceramic fibre came into use. These transformed furnace design and greatly reduced heating times and fuel consumption on batch furnaces. These materials are now in common use and are available for a temperature duty of up to 1600 °C. The thermal capacity is one tenth that of a solid refractory and the thermal conductivity is about the same as insulating brick. In fuel efficiency terms there is still some scope for further advance from lightweight refractories. A specimen computer prediction for a batch heat treatment furnace, assuming the use of a zero conductivity lining of zero weight, indicated additional fuel saving of 15% compared to currently available lightweight furnace linings.

Oxygen enriched combustion air

Commercially pure oxygen is generally produced from energy-intensive cryogenic processes. Membrane technology is becoming available which can enrich atmospheric air to 40% of oxygen. As the oxygen level increases in the combustion air, so the level of inert gases decreases such that flame temperatures rise from 1930 °C with 21% oxygen to 2800 °C with 100% oxygen. These high temperatures have particular application in the flame cutting of steel and in oxy-gas assisted steel melting.

Furnace performance has been measured using quite modest levels of oxygen enrichment. In one case enrichment to 24% oxygen led to a reduction of 10% in specific fuel consumption. On another furnace, enrichment to 22% oxygen increased output by 5% and reduced specific fuel consumption by 8%. The conclusions were that modest levels of oxygen enrichment could increase output from existing plant for a relatively low capital expenditure. For new furnaces it was more economic to build a furnace without enrichment.

Advanced control systems

Advanced sensors for measuring gas flow, temperature and concentration and advanced actuators are being developed outside the fuel industries, often for use in consumer products such as vehicles. This provides a source of low cost, high performance components of high reliability which can be engineered into control systems for fuel-fired plant. The benefits are lower fuel consumption and improved environmental performance.

For instance, if a furnace with an operating temperature of 1250 °C has an air–fuel ratio set to 10% excess air instead of the stoichiometric ratio, it will consume 20% extra fuel. If the actual temperature is really 1300°C instead of the correct 1250 °C the penalty in extra fuel is also 20%. An imprecise control system can drift within these levels of tolerance thus reducing the efficiency of the furnace.

A low cost control system has been developed for high temperature plant. Incorporating oxygen sensing in the flue gases and electronic air–fuel ratio control, it has a high turn-down capability. On a kiln producing ceramic ware operating at up to 1200 °C, fuel savings of 15% were achieved.

(b) *Analysis of the potential to increase the thermal efficiency of high temperature process plant*

So far this section has briefly described the types of furnaces used to heat materials to high temperatures and the technologies which can be used to raise thermal efficiencies. Other technologies may emerge but it can be shown that evolutionary development of existing technology can enable the limits of efficiency to be approached. Computer models have been developed for this type of plant and these have been used to explore the realistic limits to thermal efficiency.

The data (Table 8.8) show that furnace efficiency has generally increased quite markedly from 1970 to 1990 with several examples where it more than doubled. Further advances are predicted in the period up to 2010 with efficiencies reaching 75%. In some cases the limiting efficiencies are in the range 80–90%. The rate of progress and selection of technology will depend upon capital cost, fuel prices, the rate of industrial restructuring and the propensity to invest.

Table 8.8 The efficiency of oil and natural gas use in industrial processes

Furnace type /mode of operation	*Efficiency (% net) – best practice*				*Issue*
	1970	*1990*	*2010*	*Limit*	
Batch furnaces:	For batch furnaces without heat recovery to the stock, there is an efficiency limit of 80% determined by the maximum flue gas heat recoverable by air and gas preheating; the limit for light weight insulation has been closely approached with ceramic fibre				
Forging (e.g. steel 1250 ° C) heating/discharging	15	35	55	80	Light weight insulation led to large efficiency improvements in 1970s. Flue gas heat recovery will be important for future improvements
Pottery kilns programmed heating/ soaking/cooling	15	25	30	80	Energy cost is small proportion of product value. Future developments are aimed at improving quality
Melting (e.g. aluminium at 600 ° C) melting/holding/discharging	15	55	55	80	There are a large number ot furnaces still to be improved by better insulation, flue gas heat recovery, load preheating, better controls. Quality and reliability will take priority over increasing best practice efficiences in the future
Continuous furnaces:	Heat recovery by stock increases the theoretical limit for efficiency to 90% allowing for some losses through the insulation				
Steel reheating for rolling (1200 ° C)	50	50	70	90	Overcapacity in the steel industry has offset efficiency improvements made since 1970 by improved controls, flue gas heat recovery and optimized design. Better furnace utilization will be important in future together with changing factory practices (e.g. reheating hot material from the upstream continuous casting process)
Heat treatment (e.g. ferrous 900 ° C)	20	35	55	90	Lightweight insulation, flue gas heat recovery, improved furnace design and control all contribute, product quality important

Table 8.8 *Cont'd.*

Furnace type /mode of operation	*Efficiency (% net) – best practice*				*Issue*
	1970	*1990*	*2010*	*Limit*	
Pottery kiln heating/soaking/cooling	15	30	35	90	New kiln designs are aimed at improving flexibility through fast firing and improved quality. A switch from electric to natural gas will occur provided quality can be maintained
Glass melting furnace heating/melting	40	60	75	90	Flue gas heat recovery has long been practised on glass furnaces. Load preheating is necessary to achieve further significant improvements
Oxy-gas glass melting heating/melting	–	70	75	90	The incentive for oxy-gas firing will mainly be to reduce emission of NO_x
Novel advanced furnaces:	The incentive for developing novel furnaces is usually to improve quality, flexibility or reduced capital cost. Efficiency improvements are often of secondary importance.				
Rapid heaters (e.g. steel 125 ° C) heating/discharging	25	55	70	90	Technology emerged in 1970s. An advanced design could incorporate flue gas recuperation, low heat loss insulation, minimized loss to skid rails. Rapid heating offers improved product quality
Advanced glass melter heating/melting	–	–	55	90	This concept offers reduced capital cost
Permeable wall batch furnace heating/discharging	–	–	80	80	Permeable wall gives the furnace heat-up time in seconds giving improved flexibility and product quality

Batch furnaces have made significant progress since 1970 with generally a doubling of thermal efficiency. Improvements have come from lightweight furnace linings, heat recovery from flue gases into combustion air and better controls. The limit on lightweight insulation is already being approached by ceramic fibre linings but there is still scope for some further advancement. This, combined with more effective heat recovery and controls, could lead to a further doubling of efficiency. In many cases the energy cost is only a small proportion of production costs and therefore is not of high priority.

Continuous furnaces have also attained higher efficiencies since 1970 but more typically by about 50%. The effectiveness of energy efficiency measures is more dependent on a continuous high throughput and this is rarely achieved during periods of overcapacity. Further improvements of up to 30% are predicted by 2010. The limit on furnace performance is higher than for batch furnaces at 90%. This could be achievable but would depend on continuous high throughput.

Advanced furnaces relate to novel designs which seek intimate, high speed convective contact between combustion gases and the material being heated. This can lead to dramatic reductions in furnace size and instantaneous response to changes in throughput. In the advanced glass melter, glass particles are introduced into the stream of combusting gases. In the rapid heaters high velocity jets of combustion gases impinge or scrub along the surface of the stock. These types of unit can offer convenience in installation into automated production lines and flexibility in production. They could achieve limiting efficiencies of 80–90% and, in practice, this may be more readily achieved than with the more conventional designs.

Changes in production practices should also save furnace fuel. For example, in modern steel rolling mills, hot steel from the continuous casting process enters the reheating furnace at 600–800°C. This essentially halves the energy required for reheating. In modern kilns, advanced firing techniques are reducing the need for kiln furniture to support the ware, thus reducing the energy input. From the prediction, together with an awareness of the changes in industry, it should be possible to reduce specific fuel consumption in high temperature processes by 50% in the longer term.

8.3.5 The impact on world energy demand of achieving the realizable limits of efficiency for industrial processes

An assessment of the impact of achieving the limits of efficiency for industrial processes on world consumption cannot be detailed. What has been shown is that the UK industrial oil and gas consumption could be broadly halved while maintaining the same industrial output. This figure should be conservative if industries restructure and more intensive production and energy conscious methods are used.

UK energy practices probably compare favourably with most OECD countries and could be better than the CIS, Eastern Europe and the emerging economies. It is possible, therefore, that there is potential to more than halve the world oil and gas energy consumption in industry at current levels of output. With 40% of world oil and gas being used in industry, this would reduce world consumption by 20%. This would have a major positive impact on energy reserves and emissions. Should the emerging economies grow to similar energy per capita levels in industry as the OECD, then energy consumption would rise even if the limits of efficiency were attained.

8.3.6 Conclusions - industrial energy use

1. Energy consumption in both the low and high temperature sectors of industry could be reduced by about one half, while maintaining production levels.
2. Evolutionary development of existing technology could enable the limits of efficiency to be approached.
3. Advances will come from a combination of high efficiency energy-using equipment, the rationalization of production lines and the substitution of low energy materials and practices.
4. The industrial sector is expected to grow world-wide. The impact of this potential growth will outweigh the energy savings such that energy use will increase.

8.4 ENERGY USE IN COMMERCIAL BUILDINGS

Energy use in commercial buildings is 15 000 PJ/year, representing about 5% of world energy consumption. The developed economies have the highest consumption and the developing economies the least. In developed economies commercial buildings represent 7% of total energy requirements. As developing economies mature, then the proportion of energy used in commercial buildings should rise to that of the developed economies.

8.4.1 Energy use in the service sector in the OECD countries

Total energy use in the service sector of the OECD countries has grown in all countries from 1973 to 1988 (Figure 8.12). The United States uses 61% of OECD service sector energy. The energy needs differ geographically, with high heating demand in colder climates and a growth in air conditioning in hotter climates.

The service sector has seen a marked shift in energy use from fossil fuel to electricity between 1970 and 1988. All four major OECD groupings

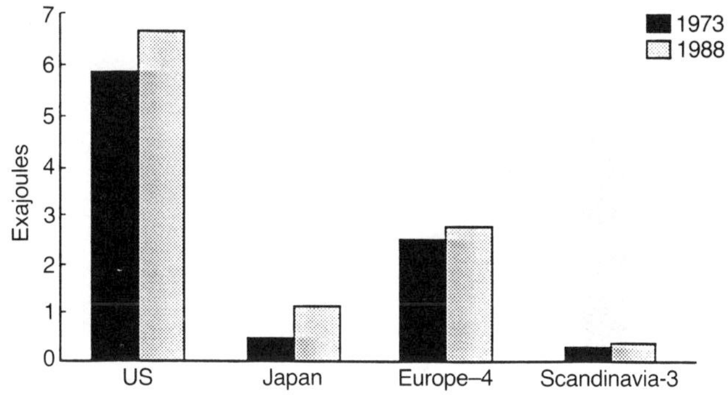

Figure 8.12 OECD-9 annual service sector energy use, 1973–1988.

increased electricity intensity by between 20% and 50% (Figure 8.13). During the same period the fuel intensity fell by between 25% and 50% (Figure 8.14). This reflected a reduction in space heating energy requirements as newer buildings with lower heating demand were constructed and energy management controls became widely used. The rise in electrical intensity is due to an increased use of business machines and air conditioning.

8.4.2 Energy saving technologies for commercial buildings

The commercial sector embraces six main categories: offices, retail, health care, education, leisure and hotels. The energy services required include space heating, hot water, catering, lighting, cooling and power for lifts, business machines, pumps, ventilation and miscellaneous services. Historically there has been a trend to reduce the space heating load while power and cooling requirements have increased as buildings improved their

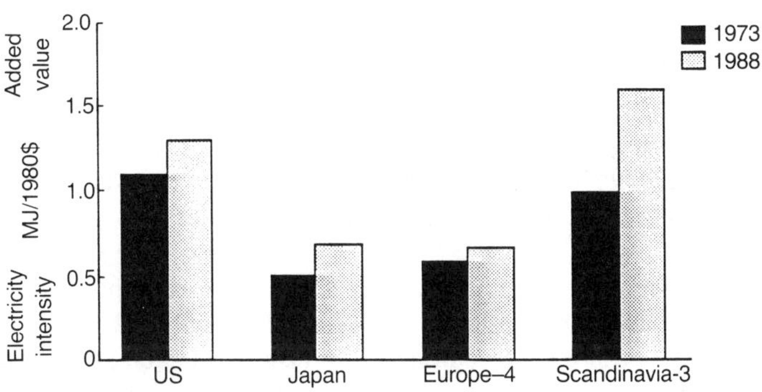

Figure 8.13 Electricity intensities for OECD-9 service sector, 1973–1988.

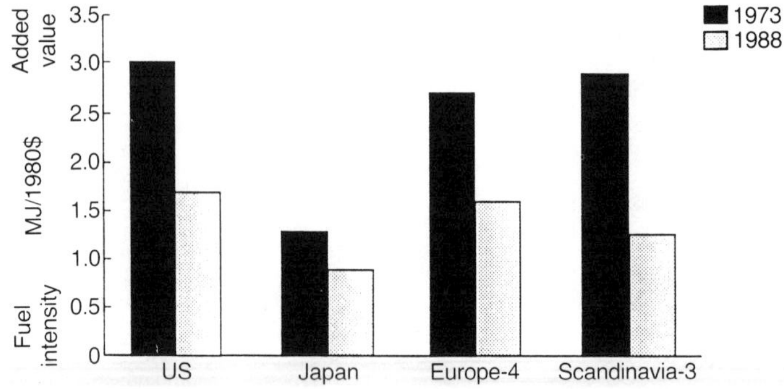

Figure 8.14 Fuel intensities for OECD-9 service sector, 1973–1988.

insulation levels and office equipment became more commonplace. At present the balance between fuel and power is about 60:40 for the United Kingdom.

The factors which will affect the future energy requirements of commercial buildings fall into four categories.

1. Improved plant and equipment services including advanced technology for heating, combined heat and power, air conditioning, lighting and business machines.
2. Advanced building designs and materials which have a major effect on the energy service requirements.
3. Increased expectations in developed and developing economies of the level and quality of facilities in the built environment.
4. Shifting patterns in the workplace from manufacturing to service industries, a trend towards working from home and an increase in leisure activities.

(a) *Heating and service hot water*

In the developed countries, around 80% of heating is provided by hydrocarbon fuel burning boilers, the remainder is mainly direct electric heating. Boiler systems have typical seasonal efficiencies of around 65% with a maximum of about 80%. Over the next few decades it is envisaged that condensing boilers will become the norm. These, together with improved controls and insulation, should give system efficiencies approaching 90%.

Fuel-fired heat pumps could give ratios of 'heat out' to 'fuel in' of two, using current technology. Developments in heat exchangers, prime movers, compressors and refrigerant technology should give further advances and ratios of four could be achievable.

Current co-generation plant can generate electrical power at around 30% efficiency with a further 50% of the fuel energy converted to useful heat. Fuel cell packages offer the potential of electrical conversion efficiencies up to 65% with perhaps an additional 20% of useful heat. As building heat demands reduce, the development of higher efficiency co-generation will become increasingly important.

In addition to active heating plant, passive techniques will have a significant influence on energy requirements. Passive solar design is one of the most important strategies for the displacement of conventional fossil fuels. Techniques have been refined over the last few decades and some central European buildings have recently been designed to take around 60% of their heating needs from the sun. Looking to the future there is no technological barrier to designing a building which takes all of its heat requirements from the sun. Development of transparent insulation materials, electrochromic glass and reactive building fabrics will all aid the development of solar buildings.

Seasonal heat storage is another passive technique which could significantly affect energy requirements of buildings. The concept is to remove and store excess heat in the summer which is then used to provide heat in the winter. Charging of the store in the summer months also provides cooling to the building. Solar collectors linked to the store provide an additional source of heat. The store medium is subsoil, rock or an aquifer.

An experimental subsoil store linked with solar collectors has been successfully developed for an office in Sweden. It provides virtually all the building's heat requirements (70 GJ/year). Heat supply costs for another scheme in Sweden are estimated at 6p/kWh, about twice the cost of using fossil fuels. It will be the relative cost of fossil fuels that determines the initial speed of uptake.

(b) *Cooling*

Fuel-fired cooling is commonly provided by absorption machines which can be directly gas fired or driven by steam. Coefficients of performance are up to 1.3 for direct-fired machines. Chlorinated fluorocarbons are not used and so the technology is environmentally friendlier. Other options are being developed to reduce capital and running costs. These include gas-engine-driven vapour compression machines and cooling systems based on desiccant technology. In the long term, coefficients of performance of 2.0 should be possible.

The potential for passive cooling techniques is also being investigated. These generally involve the generation of a cold store during night time periods, possibly involving evaporative cooling. This approach is more applicable to climates with low night time temperatures. A recent study of passive cooling has been carried out by British Gas comparing mechanical ventilation with evaporation cooling. A dynamic building energy model was constructed to simulate the performance of prototype plant and controls in a range of modern office buildings. The model incorporated details of building fabric and occupancy patterns as well as hourly weather data throughout a full year.

The evaporative cooler significantly reduces peak internal temperatures in comparison with a forced ventilation system. Comfort conditions are predicted to be acceptable throughout the year; however, on the hottest few days of the year they are not ideal. The energy used by fans is more than offset by the heat recuperated in winter by the inlet/exhaust heat exchanger which is an integral part of an evaporative cooling system. With the system providing comfort conditions close to optimum it is reasonable to assume that a combination of the technique with seasonal heat storage and/or building thermal mass would enable optimum conditions to be maintained throughout the year.

(c) *Lighting*

The luminous efficacy of modern office lighting is around 60 lumens/watt (lm/W). In the next few decades this is expected to increase to around 100 lm/W and within the 21st century should rise close to its theoretical maximum at around 250 lm/W. Energy needs for lighting in new buildings should halve in the next few years and reduce to one quarter in the long term. Reduction of lighting loads will reduce the cooling demand and increase the heating demand in winter months. However, in many buildings the heat dissipated from lighting is of similar magnitude to the cooling load. In the long term, optical fibres could be used to distribute natural day lighting around a building with minimal heat gains.

(d) *Small power and office machinery*

The main small power loads are computers, visual display units, printers and copiers. In the early 1980s it was widely assumed that small power loads would increase substantially with the growing use of business machines and desktop computers. However, newly developed equipment has comparatively low power requirements and so although the overall numbers of equipment have substantially increased, the total power consumption is roughly constant. In the long term, the development of low power liquid crystal display screens and more energy efficient processors could actually lead to a decrease in energy requirements. Additionally, the need for printers and photocopiers may well diminish as paper-free communication becomes the norm. Overall, the small power requirements of buildings are likely to remain fairly steady in the near term but reduce gradually in the long term.

(e) *Catering*

Catering is a highly traditional area and open flame cooking is likely to remain, together with a mix of electrical means. Overall energy use is expected to remain similar, with the lower energy demands of convenience food preparations being offset by the increase in dining out. Waste heat from kitchens will probably be recuperated at some stage in the future.

8.4.3 Building design and materials

The building structure can be used as a moderator of heating and cooling demands. Initial emphasis is on reducing cooling requirements, because costs can be offset against the high cost of conventional cooling systems (up to £200/m^2). The use of natural ventilation to reduce cooling requirements is being actively pursued in a number of countries. Atria and thermal stacks are being developed to enhance the air change rates on design days when dry air temperatures reach a maximum and wind speeds are low or zero. Both exploit natural buoyancy effects to create pressure differentials around the building, to provide the driving force for natural ventilation. Modelling work has shown the potential to maintain acceptable comfort conditions in Northern European climates without resorting to active cooling.

Another technique is to use the thermal mass of the building fabric to help moderate the temperature swings within a building. In summer the heat gains within a building can, in part, be absorbed by the structure. The structure is then flushed overnight by cooler outside air. The technique works equally well in the winter months when heat from solar gains, lighting, equipment and occupants is absorbed by the structure during the daytime and helps to offset heat requirements over night. The technique is becoming popular with one commercial organization already responsible for over 1 million square metres of commercial building in Europe.

In the long term it is probable that the active heating demands of commercial buildings can be virtually eliminated. Some office buildings in Sweden which integrate thermal mass with extremely high levels of thermal insulation can 'coast' through periods of several days when the outside temperature remains sub-zero. With future refinements to the technique and improvements in insulation materials, the active heating requirements seem set to become minimal in all but extreme climates and conditions of use.

(a) *Developing countries*

Most of the work on passive building design has been applied to buildings in Northern Europe and North America, often in locations where summer cooling could be thought of more as a luxury than an essential. Information concerning the application of these techniques to warmer climates is very limited although clearly important in the long term. Although building design and new materials could greatly reduce the cooling requirements of commercial buildings in hot climates, they are unlikely to offer fully acceptable comfort conditions.

8.4.4 Energy use of future buildings

By taking account of the potential advances in plant and design discussed above, an estimate has been made of the energy requirements of future buildings. Offices and hotels have been selected for the evaluation as these represent the lower and upper levels of energy intensity in commercial buildings. Data have been calculated for the medium-term future (20 years) and the long term (conceivable). For comparative purposes data are also given for contemporary buildings (typical and best practice) (Figures 8.15 and 8.16). In 20 years' time the energy requirements of a new-build building could be around 25% of typical current stock. In the long term, overall energy requirements of new buildings are expected to fall to between 10% and 20% of current levels.

There will also be a shift in the balance between fuel and electricity demand. In hotels the fuel: power ratio will move from the current 65:35 to 50:50 (Figure 8.17). In offices the ratio will move from 60:40 to 40:60 (Figure 8.18). These ratios could be close to the ideal for combined heat and power systems which could become the favoured energy system option in the longer term.

The relative importance of the end-use categories will alter significantly. Fans, pumps, office equipment, service hot water, catering and small power will all become increasingly significant. Between them these areas could

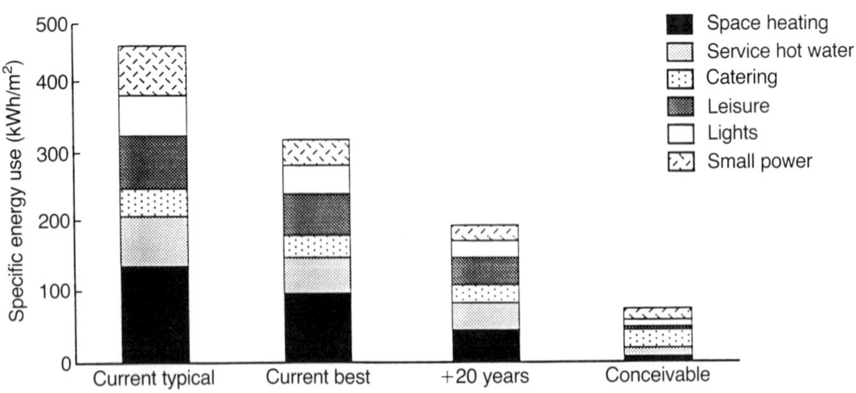

Figure 8.15 Energy use in UK hotels.

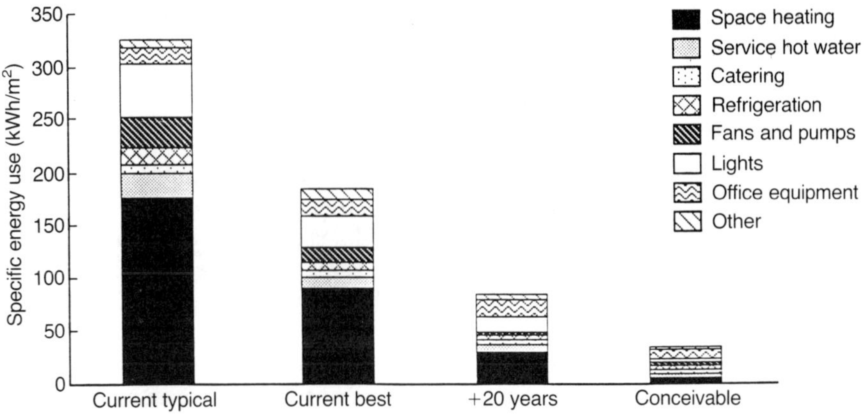

Figure 8.16 Energy use in UK offices.

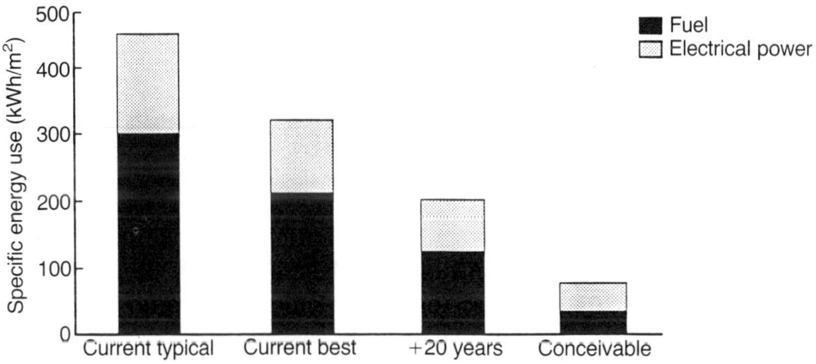

Figure 8.17 Energy use in UK hotels.

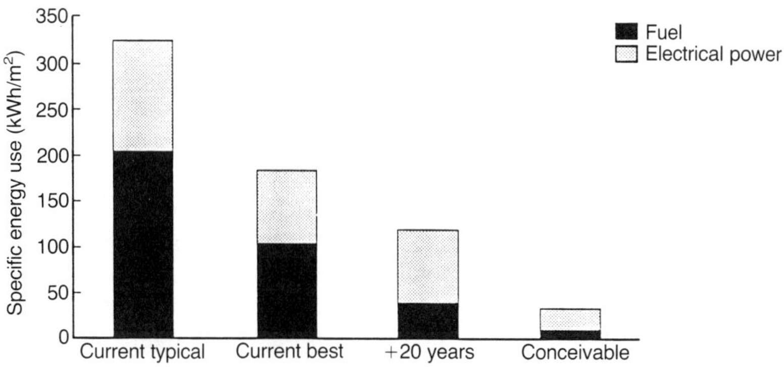

Figure 8.18 Energy use in UK offices.

account for almost 60% of a building's needs – this compares to around 40% in current buildings.

8.4.5 Industrial buildings

Industrial buildings range from corrugated steel sheet open structures designed to keep out the extremes of weather to highly insulated well-designed accommodation. In the larger buildings a local heater may be used to provide a point source of heating for occupants, while the bulk of the building is unheated. In modern high technology factories comfort levels are comparable to office accommodation.

Heating of factory buildings is by steam heated panels, hot water radiators, overhead radiant plaques and radiant tubes, or warm air heaters. The trend is towards radiant and warm air systems which give efficiency advantages. Extensive work has been carried out in the United Kingdom on modern factory designs to develop buildings and heating systems with low energy requirements and high levels of comfort. Building insulation and air tightness are close to optimum for the United Kingdom for the low energy factories and the gas-fired heaters have efficiencies of 80%. A key point is the provision of high comfort levels. With good system design, temperature variations of less than 1 °C can be achieved. To reduce energy requirements further, advanced energy systems such as heat pumps and co-generation would be needed.

8.4.6 The impact on world energy demand of achieving the realizable limits of efficiency for commercial building

On a global basis, the demand for energy will be greatly influenced by future changes in population and economic prosperity. By combining projections of future populations with the above estimates of energy use in new buildings, an estimate of world-wide energy use in commercial buildings has been made (Figure 8.19).

The prediction indicates a 6% decline in energy use over the next 20 years. This is mainly due to energy efficient new buildings replacing older, less efficient stock in the OECD countries. In the long term, however, the energy use could rise to around 10% higher than current levels. This is mainly attributable to the assumed growth in world population combined with the economic development of the non-OECD countries to OECD levels of energy consumption per capita. These factors are likely to more than offset the energy savings made possible through efficiency improvements.

The estimate takes no account of other factors which could impact on future energy use. For example, there are expected to be significant changes in the relative importance of some of the categories of commercial building. Additionally, improving information technology and communications will

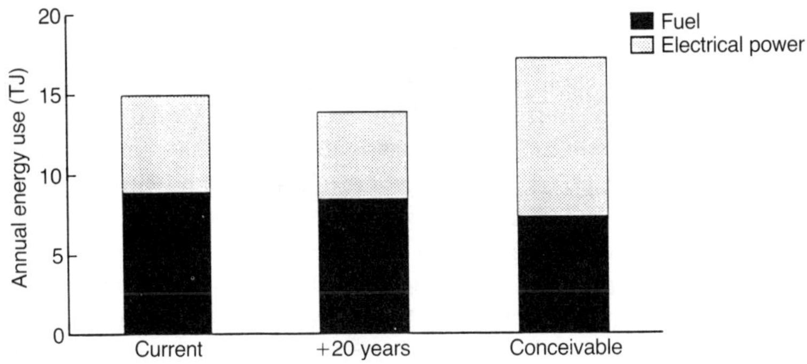

Figure 8.19 Energy use in commercial buildings, world-wide estimate.

provide greater opportunity to work from home. Also the leisure, retailing and hotel sectors have been expanding rapidly in comparison with other areas, largely due to the increasing standard of living and available leisure time of individuals.

8.4.7 Conclusions - commercial building energy use

1. Energy consumption in new commercial buildings could be reduced to between 10% and 20% of current typical values.
2. Advances will come from a combination of high efficiency energy-using equipment and building designs which make full use of possible energy saving technology.
3. The commercial sector is expected to grow world-wide. The impact of this growth will outweigh the potential savings such that energy use in commercial buildings will increase slightly.

8.5 DISCUSSION

This chapter gives some indication of the limits of efficiency for oil and gas in power generation, industry and commerce. Studies on transportation and residential use are in progress. It is stressed that energy consumption is just as dependent on industrial practices and building design as on the efficiency of the fuel consuming component in the system.

The lifetime for reserves can be extended by increasing the efficiency of utilization. There is, however, a gap between technical capability and the performance of commonly used technology. Any increase in the use of high efficiency equipment must be driven by economic or regulatory pressure. The consumer will not pay to save energy for altruistic reasons. None the less, during the timespan of a millennium it is reasonable to assume that the technical limits of efficiency will be approached.

The proven and possible reserves of oil and gas give a lifetime of less than a century before they are depleted. An increase in world economic activity is probable and desirable but could shorten the timescale considerably. It will also increase the output of carbon dioxide. On a global and historical scale the issue of depletion of reserves is urgent. At the commercial and operational level, supplies of energy are cheap and plentiful. There is therefore a danger that commercially sensible short-term decisions could jeopardize the strategic longer term future.

The growth in the use of energy is likely to increase the rate of output of carbon dioxide into the future. The effects of this are unknown but, on balance, are thought to be deleterious to the general well-being of life on Earth. Even if the rate of output is reduced by political and practical action, this will only alter the time by which all fossil fuels are consumed and their carbon content is released as carbon dioxide. Unless the absorption mechanisms for carbon dioxide are reinforced naturally or by specific action then the carbon dioxide level will rise to a predetermined level.

A continuing increase in the efficiency of fossil fuel use will help extend the lifetime of reserves and reduce the rate of output of carbon dioxide. Its encouragement would seem to be a sensible route. It will give more time to ascertain precisely the implications of an increase in carbon dioxide in the Earth's atmosphere and to develop alternative energy strategies. The development of high efficiency plant and practices and their widespread use when commercially available should be encouraged in both the developed and developing economies.

8.6 GENERAL CONCLUSIONS

1. The world demand for energy is likely to rise in all countries with a faster rate of growth in the emerging economies.
2. Evolutionary development of existing technology could enable the limits of efficiency to be approached. Other more attractive technologies may emerge but are unlikely to offer significantly higher efficiencies.
3. An increase in world economic activity is probable and desirable if large sections of the world community are to enjoy the same level of amenity as the industrialized countries. This growth is likely to increase energy consumption even if the efficiency of energy use continues to rise and the limits of efficiency are approached.
4. Unless reserves increase markedly, the widespread use of the most efficient gas- and oil-fired technology would not be sufficient on its own to meet the world's aspirations for energy in the longer term. Other environmentally benign and safe energy options will need to be developed and applied during the 21st century.
5. The global output of carbon dioxide from fossil fuel use is likely to rise.

6. Gas and oil can both be used in most applications. The selection will depend on cost, reliability, maintenance needs, emissions and convenience.

ACKNOWLEDGEMENTS

The author would like to thank British Gas plc for permission to publish this chapter, and his colleagues at the Gas Research Centre for contributing to and reviewing it. They include the following: Chris Hodrien, David Haynes, Bob Tucker, Ian Freeman, John Chapman, Bob Carpenter, Geoff Parkinson, John Margetts, Debbie Brown, Neil Gabriels, Frank Doherty, Neil Slater, Alison Newsome, Graham Goodfellow, Pat Allen, Pat Moore and Malcolm Hoggarth.

FURTHER READING

General

BP (British Petroleum) *Statistical Review of World Energy*.
Digest of United Kingdom Energy Statistics (DUKES), Department of Energy, HMSO, London
Schipper, L. and Meyers, S. (1992) *Energy Efficiency and Human Activity: Past Trends and Future Prospects*. Cambridge Energy Studies.
World Bank (1990) *World Development Report*. World Bank, Washington DC.
World Energy Council (1993) *Energy for Tomorrow's World*. St Martin's Press, New York.

Power generation

Janes, J. (1992) California Energy Commission - the optimally aggregated and chemically recuperated gas turbine power plant: a singular solution. Energy Daily Conference, Washington, USA, 26 June.
Kalina, A.I. *et al.* (1991) Further technical aspects and economics of a utility-size Kalina bottoming cycle. *ASME Report 91-GT-365*. (Presented at the International Gas Turbine and Aeroengine Congress and Exposition, Orlando, Fl, June 1991.)
Masters, J. and Dicks, A.L. (1989) The potential and market penetration of molten carbonate fuel cells. Paper presented at International Gas Union 18th World Gas Conference, Berlin, 9 July 1989.
Masters, J., Pugh, R.R. and Weller, G.B. (1991) Combined Heat and Power - generating new markets. Paper presented to the Scottish Section of the Institution of Gas Engineers, Glasgow, 13 February 1991. (MRS E 582)
Nakhamkin, M. *et al. CAES Plant with Steam Generation*: *Preliminary Design and Cost Analysis*.
Ribesse, J. The isotherm partial oxidation gas turbine. *European Journal* M, **36**, (1), 27–32.
Turbine Tech (1993) Reports from gas turbine manufacturers on their latest technologies and applications. *Diesel & Gas Turbine Worldwide*, October, 18–24.

Industrial energy use

Caddet, Cash. (1992) *Process Integration - International Experiences and Future Opportunities*, Gothenburg.

Clatworthy, E. and Masters, J. (1988) Displacement of steam by direct firing in industry and commerce. Paper presented at IGU 17th World Gas Conference, Washington DC, 5-9 June 1988.

Davies, R.M. and Masters, J. Emissions from combustion plant and processes industrial and large commercial markets. Paper presented at IGE Gas Engineering and the environment course, Oxford, March 1990. (MRS E 590)

Energy Consumption Guide No. 9 (1991) *Continuous Steel Reheating Furnaces*. EEO, February.

Energy Efficiency Office (1984) *Energy Use and Energy Efficiency in UK Manufacturing Industry up to the Year 2000*, Volume 1. Energy Efficiency Series, HMSO, London.

Fricker, N.J., Hoggarth, M.L. and Masters, J. (1981) Efficiency Use of Energy in High Temperature Furnaces, 47th IGE Autumn Meeting, 24–25 November 1981, IGE Communication 1160.

Jones, D.A. (1992) The integration of combined heat and power for process heating applications. Paper given to the International Gas Research Conference, Orlando, November 1992.

Linnhoff, B. (1993) Pinch analysis - a state-of-the-art overview. *Trans. IChemE*, **71**, Part A, September.

Masters, J., Pearson, J. and Read, M.A. (1980) Opportunities for Gas Engine Driven Heat Pumps in the Industrial and Commercial Markets, 46th IGE Autumn Meeting, 11–12 November 1980, IGE Communication 1129.

Masters, J. and Webb, R.J. (1983) The development of a recuperative burner for gas fired furnaces. Paper presented for Royal Society Esso Energy Award, 2 November 1983. (MRS E 412)

Pomfret, K.F. and Waddington, J.D. (1992) The development and exploitation of gas-fired rapid heating furnaces for metal reheating. Esso Energy Award Lecture, 1990. *Phil. Trans. R. Soc., London* A, **338**, 69–82.

Ramshaw, C. (1984) Process intensification - incentives and opportunities. Paper given to Chem Read Conference 1984, Volume 1.

Rhine, J.M. and Tucker, R.J. (1991) *Modelling of Gas-Fired Furnace and Boilers*. British Gas Technical Monograph, McGraw-Hill.

Commercial buildings

Abel, E. *et al.* (1990) New technologies for heating and cooling supply in offices, CADDET analyses series no. 3, IEA, June.

Braham, D. (1992) *Better Insulated Buildings (II)*, G.D. Braham.

Crane, J.M. (1991) The Consumer Research Association new laboratory and office - a design case study, Group Bywaters & Anders, CIBSE National Conference, 7-9 April 1991.

Herring, H., Hardcastle, R. and Phillipson, R. (1989) *Energy Use and Energy Efficiency in UK Commercial and Public Buildings up to the Year 2000*. Energy Efficiency Series 6, HMSO, London.

IEA (1991) Passive and hybrid solar commercial buildings, Advanced Case Studies Seminar, IEA, 24 April.

Masters, J., Pearson, J. and Chisholm, D. (1992) Technological opportunities for gas in commercial buildings, 128th Annual General Meeting, 20–22 May 1992, IGE Communication 1459.

Masters, J., Thompson, K. and Whitman, D. (1993) Integrated heating/cooling Technology in commercial buildings. Paper presented at the IEA Conference on Natural Gas Technologies, Kyoto, November 1993. (GRC P 0012)

Olivier, D. and Andrews, S. (1989) *Energy Storage Systems - Past, Present and Future Applications*, Maclean Hunter Business Studies, UK.

Taylor, T.B. *et al.* (1984) Statewide assessment of ice ponds for process cooling, Nova Inc. Report, Gaithersburg, Maryland, USA.

9

Transport: propulsion into the 21st century

P.J. Hills

9.1 INTRODUCTION

Transport in the United Kingdom during the latter half of the 20th century has been dominated by the growth of private car ownership. All the attendant problems of congestion, pollution and accidents are conspiring to undermine the huge personal and social benefits that individuals and households derive from car ownership and use. Nor has this process run its full course; the Department of Transport in its National Road Traffic Forecasts (NRTF) foresees the possibility of the total annual vehicle-km more than doubling between now and the year 2030, before some kind of 'saturation' in car use might be reached. The extent to which these forecasts are reflecting the powerful economic forces that determine future traffic demand or are themselves driven by current transport policy is the subject of much debate. More radical policies, such as tolling inter-urban motorways, congestion pricing and traffic calming in urban areas and encouraging the spread of tele-commuting could modify and reshape future travel demands substantially, with significant implications for daily life.

In this chapter, mention will be made of the forces which determine traffic growth and of the current prospects for fuel conservation, by applying known technology to conventional i.c. engines. The limitations of the challenges offered by less conventional engines will be discussed, as well as those of alternative fuels. A glimpse will be offered of current European developments in automatic toll-collection/road-use pricing systems, which could be used to restrain traffic demand and reduce the costs of congestion. The chapter concludes with a longer term look at the prospects for reducing exhaust emissions and the implications of these changes in the UK context and world-wide.

9.2 THE FORECAST GROWTH IN UK TRAFFIC DEMAND

The amount of travel in the United Kingdom has been growing almost without interruption since the end of the Second World War. From a total of about 200×10^9 person-km in 1952, it has expanded almost fourfold to 760×10^9 person-km in 1992. Proportionately, the fastest growth has been in airline travel but from an exceedingly low base (2% of travel in 1952). Next comes car use, with a more than 10-fold increase but from a larger base (25% of travel in 1952). Although in absolute numbers travel by train has held its own, proportionately it has declined (from 20% in 1952 to about 5% now). Bus and cycle use have fared even worse, declining both absolutely and proportionately. Since bus deregulation in 1986, for example, roughly one-third of all bus passengers have deserted the buses and switched to other modes (mainly cars).

Everywhere else besides the central and inner areas of our larger cities, the private car is now the dominant mode of transport. In smaller towns and rural areas, it is increasingly the only form of personal transport. To the extent that we can be said to have a 'transport problem' in this country, it is essentially a car problem. Nor have we seen its full extent. For, in terms of car ownership per head, the United Kingdom is at or rather below the average for all Western European countries and way below the ownership levels in the United States, Canada, Australia and New Zealand. The car ownership rates for OECD countries over the last 40 years are shown in Figure 9.1.

Several important implications are revealed in Figure 9.1, which have a bearing on future energy requirements and prospects for global pollution. These are:

1. mass car ownership is a privilege afforded (at the moment) only to the rich nations (i.e. the curves for non-OECD countries all lie below that of Greece);
2. the growth of car ownership is 'driven' by real income per head (if all the curves shown in Figure 9.1 were normalized, in relation to national GDP per head, they would virtually coincide);
3. if there is a meaningful 'saturation' level of car ownership, it is probably in excess of the US level of 580 cars per 1000 inhabitants;
4. there is no obvious reason why Europeans (at the personal level) will not aspire to North American levels of car ownership and use, as a direct consequence of rising real incomes over the next 25–30 years.

More ominously, in relation to world energy and global pollution, there is no corresponding reason why the vast bulk of non-OECD populations will not aspire to car-owning, just as soon as they can afford it. After all, more than two thirds of the world's population now live in countries whose average car ownership is below 100 cars per 1000 inhabitants.

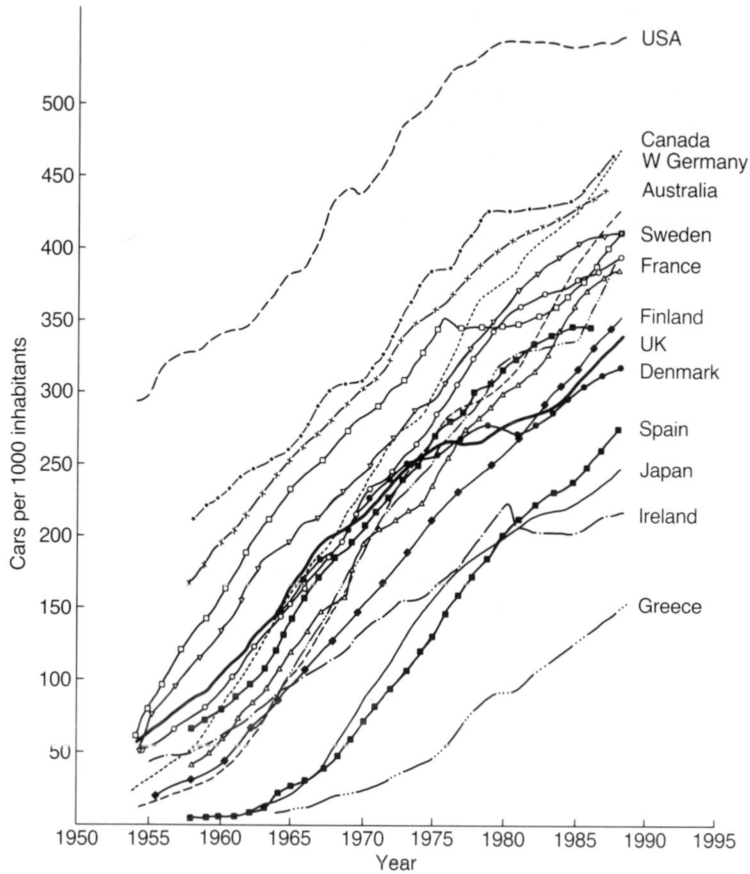

Figure 9.1 Growth in car ownership in a number of OECD countries. (Source: International Road Federation Statistics.)

Looking at the United Kingdom's experience in comparison to other European countries, the period 1975–1982 appears to have been one of relatively low growth. This was, straightforwardly, a reflection of our relatively poor economic performance. The main causes of low traffic growth during this late 1970s and early 1980s period were:

1. growth of real income per head averaged only 1–1.5% per annum;
2. car prices were kept artificially high due to the EEC car-distributors' 'cartel';
3. a special 10% tax (repealed in 1992) was imposed on new cars over and above VAT;
4. a widespread misperception prevailed concerning the real price of fuel (now returned to pre-OPEC levels);

5. there were mounting subsidies to public transport (bus grant revenue support PSO, etc.), in the years up to deregulation in 1986.

The relatively poor performance of the UK economy during these years can be seen clearly in Figure 9.1, such that by 1982, the (then) West Germany had a car ownership level which exceeded that of the United Kingdom by nearly 50%.

The late 1980s, however, saw a strong surge (with more than 2.0 million new cars per year being sold in Britain for several consecutive years). With hindsight, the causes of the surge in car ownership during the late 1980s were:

1. growth of real incomes per head was averaging more than 3.5% per annum;
2. bus deregulation (in all areas outside Greater London) led to withdrawal of public subsidies;
3. decentralization of land uses increased, due to loosening of planning controls and an accompanying building boom.

The recession (1990–1993) has levelled off car ownership in the United Kingdom as in most other European countries. However, despite the effect of the recession, the current (1994) level of car ownership is still close to the 'high growth' forecast of the Department of Transport's NRTF in 1987. Moreover, as economic growth returns, a further surge in car ownership and use can be expected. The conditions for this to happen are all in place, namely:

1. a sustained growth in real incomes per head > 3% per annum would mark the end of recession;
2. a slow decentralization of land use activities is continuing;
3. regional catchments for new out-of-town retail and leisure centres are being established;
4. there is a steady reduction in residential density and growing pressure on Green Belts;
5. price wars could erupt soon, due to overproduction in world car manufacture;
6. the steady contraction of public transport continues in the absence of subsidies or revenue support.

If most or all of these trends, already clearly evident, continue, then the car population in the United Kingdom could rise from 22 million now to as much as 36 million by the year 2030. Even at this level, car ownership would barely have reached that of the United States today. Whether a 'saturation' level has been reached in parts of the United States is arguable, with a surfeit of cars in relation to drivers and a declining mileage per car. It is also arguable whether Europeans, with more compact towns, higher residential densities and better public transport will emulate the US experience. In any event it is clear that the official 'high growth' projection of the NRTF is not unrealistic. These forecasts imply a growth over 1987 traffic levels of between + 82% and + 143% by 2025.

When we speak of mass car ownership, it seems to imply that everyone who wishes will own a car or have access to one. On present trends, even by 2010, about a quarter of all households in the United Kingdom will have no car available. Among those who will have, the balance between one-car-owning and multi-car-owning households will change steadily. Thus, towards the end of this decade, more new cars will be bought by *already* car-owning households than by *non*-car-owning households. As saturation approaches (say, by 2025), a large minority of the population (probably 20% at least) will not have a private car available and they will remain dependent on such public transport as has survived.

9.3 EMERGING POLICIES FOR ROAD TRANSPORT

The implications for public transport in the United Kingdom, outside the very few large cities, of this prospective onward march of private transport are serious, especially for the 12 million or more people dependent upon it. For, without major intervention to sustain transport services, either by public investment or revenue support (or both) over the coming years, public transport is likely to fall victim progressively to the 'unfair' challenge posed by widespread car use. Not only are buses, coaches and (to a lesser extent) trains seen to offer an inferior level of service for most journeys, but their cost-base will tend to rise as patronage declines, while private transport costs could continue to fall. Meanwhile, the external costs of car use in the form of congestion, pollution and accidents are borne by the community as a whole.

Signs are appearing in policy statements emanating from Brussels and from various Member States of the EU that this basic flow is understood. The remedy is seen to lie in most cases not with a return to indiscriminate subsidies for public transport but with a move to more direct (variable) charging for private car travel 'at the point of use'.

Emerging policies in Europe on road traffic are based upon five key points.

1. Growth in traffic is potentially greater than the realistic expansion of infrastructural capacity to accommodate it (even if this were universally accepted as the best policy).
2. The investment cost of new and improved infrastructure is seen as an increasing burden on public sector budgets – more private capital and joint ventures are to be encouraged.

3. Overall limits will be necessary on traffic growth and a fuel consumption to meet increasingly stringent environmental targets.
4. There is a widespread adoption, as in other areas of the economy, of the 'users should pay' principle.
5. Charges for road use do not properly reflect the marginal social costs of pollution, congestion and accidents. In future, they should.

9.4 CURRENT RESEARCH IN EUROPE ON ROAD USE PRICING

These emerging policies all point to the need for radical changes in the taxation of vehicles and in the pricing for the use they make of the road infrastructure. The more likely changes will include the following.

1. Substantial increases in fuel tax, to compensate for the environmental costs of pollution.
2. Reduction in the annual vehicle excise duty (VED), to shift the balance of charging from 'fixed' to variable.
3. The introduction of automatic tolling on motorways and congestion-pricing in city centres. It is expected that, with strongly rising demand for road use and a generally low price-elasticity of that demand, road use pricing will generate substantial net revenues. These would more than offset the reduction in VED and (in principle, at least) provide a major source of finance for reinvestment in infrastructure, environmental improvements and even support for public transport.

Indeed, unlike almost all other transport policy 'instruments', road use pricing would address the emerging policy needs across a wide front. Specifically, there are four distinct (but overlapping) objectives that can be met by introducing a 'pay as you go' policy.

1. raise revenue to finance further intrastructural investment (e.g. toll-rings in Norway);
2. limit traffic growth overall to a predetermined level for environmental or operational reasons (e.g. *Rekning Reiden* project in The Netherlands);
3. ration the allocation of scarce roadspace between competing claimants (e.g. differential licensing in Singapore);
4. 'internalize' the marginal social costs of congestion in the (perceived) price paid by vehicle users levied at the point of use (e.g. congestion-pricing proposal for Cambridge).

In practice, the success of all these different applications of road use pricing hinges on a reliable, high-speed, two-way communication between a roadside 'tolling point' and a single in-vehicle unit designed to serve all the applications (and perhaps others, like traffic information and route guidance). Moreover, tolling must be achieved without any interruption of

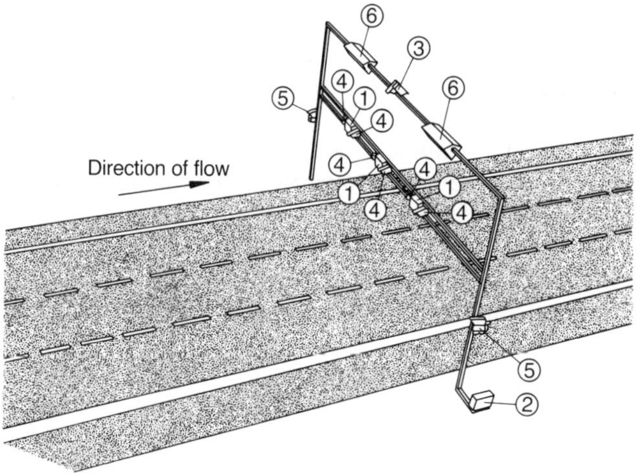

Figure 9.2 Automatic toll collection at speed, with full enforcement, using a single gantry: (1) infra-red vehicle detectors; (2) gantry controller; (3) transponder range-finding device; (4) video cameras for enforcement; (5) transponder localization cameras; (6) 'roadside'; beacons (transceivers).

traffic flow. In the current motorway-tolling demonstration in the United Kingdom, for example, the DoT has made it clear that only a single gantry can be used to span all lanes of a motorway (Figure 9.2). Building huge toll-plazas for 'stop and pay' transactions is unrealistic.

Fortunately, the technology for automatic debiting systems based upon a transponder/smartcard in-vehicle device has developed strongly over the last five years in Europe, notably in the EC project ADEPT (Automatic Debiting and Electronic Payment for Transport). The ADEPT project is an EC-funded research project within the DRIVE-ATT (Advanced Transport Telematics) Programme. ADEPT is headed by the Transport Operations Research Group (TORG) at the University of Newcastle upon Tyne and comprises 16 organizations (industrial companies, research organizations and infrastructure owners) from eight European countries. The objectives of the ADEPT project are to develop a two-way communication system between the roadside and vehicles (which may be applied to automatic non-stop tolling, road-pricing or charging for the use of car parks) and to implement the system in coordinated field trials in five cities across Europe.

The heart of the ADEPT system is a microwave two-way communication link between the roadside and a moving vehicle, which may be used on single-lane roads or on busy multi-lane motorways and allows the system at the roadside to communicate with many vehicles at once. The high speed communication is used to perform an automatic debiting transaction, whereby the vehicle's transponder has credit units deducted from its smart card as it passes a roadside charging point. This is achieved at full highway speeds, without the need for the vehicle to slow down or for the user to perform any action.

In the vehicle, the equipment consists of a small box attached to the windscreen, which contains the microwave communications transponder and a computer (Figure 9.3). Additionally, this box is interfaced to a smart card (which is used to carry the user's electronic credit) and to a keyboard/display panel, which is attached to the dashboard. At the roadside, the system consists of communications beacons mounted on a roadside post or an a gantry above the roadway. These beacons are controlled by a local or central computer, which also controls the on-line video enforcement system and provides for connections to traffic control centres and an integrated payment network. The system may also be used to monitor traffic on the road network and to provide real-time guidance and traffic information directly to the driver.

The ADEPT system offers the flexibility to operate an automatic debiting system in many ways, whether for the charging of a fixed fee for the use of a toll road or car park or whether it is used to calculate a variable road use charge based upon certain parameters – such as time of day, prevailing traffic conditions, time elapsed or distance travelled – to name but a few. Five field trials of the ADEPT system are scheduled to start before the end of 1994, as follows.

1. Göteborg (Sweden) – multi-lane road-use pricing and in-vehicle driver information (November 1992);
2. Cambridge (United Kingdom) – congestion charging (October 1993);
3. Thessaloniki (Greece) – mono-lane and multi-lane toll collection (February 1994);
4. Lisbon (Portugal) – car-park booking, guidance and automatic fee collection (May 1994);
5. Trondheim (Norway) – integrated payment and the use of multi-service smart cards (November 1994).

The demonstration of the ADEPT system at the Cambridge test site is the first demonstration of congestion metering, and indeed road use pricing technology, in the United Kingdom. This system differs somewhat in operation from the other variants of the ADEPT system to be implemented in the other field trials, as it requires the additional input of a distance-measuring sensor in the vehicle.

The Cambridge experiment of congestion metering operates by using microwave beacons to 'switch on' the in-vehicle equipment as it enters a predefined cordon around the City of Cambridge. Once switched on, the distance sensor provides information to the in-vehicle system, to enable the time taken to travel a unit distance to be calculated. If this time is greater than a predetermined value, it is presumed that the vehicle is travelling on a congested road and a charge is deducted accordingly.

Electronic credit is deducted from the user's smartcard. If the vehicle is travelling on an uncongested road, on the other hand, then no charge is deducted. This process continues until the vehicle leaves the cordon area and the in-vehicle equipment is 'switched off' via a roadside exit beacon. No charging may occur outside the cordon.

The in-vehicle unit comprises the ADEPT transponder, smartcard and keyboard/display panel. An additional input from a distance sensor is provided by a connection to a dashboard speedometer. When the speed of the vehicle is lower than a predetermined average it is assessed that the vehicle is travelling in, and contributing to, congestion. Thus, a charge is levied and the credit units being deducted from the smart card are displayed. On leaving the charging area, the meter is switched off using the roadside beacon and the total number of credit units consumed within the area is displayed.

Other elements of the Cambridge demonstration include:

1. the provision of both TMC-coded and free text driver information conveyed over the microwave link to the driver;
2. an on-line video enforcement system at the entry beacon points;
3. a PC-based smartcard service station where the facility to write personalized data and add credit to the smartcard is provided. This station may also be used to read the balance and transaction history which is stored on the smartcard;
4. the demonstration of other charging algorithms (real distance-based charging, open toll collection and closed toll collection). All these 'applications' reside together in the transponder and are automatically activated and controlled by the roadside beacon.

The first phase of the public demonstration of the ADEPT system in Cambridge is now complete. Newcastle University is currently undertaking a detailed technical evaluation of the system and the applications and an assessment of the performance and relative merits of the charging algorithms. This is funded under a separate contract awarded by the UK Department of Transport.

The system illustrated in Figure 9.3 was that which was deployed in the ADEPT field trial demonstration of congestion pricing in Cambridge during October 1993. However, essentially the same equipment will be used in the UK motorway-tolling demonstrations during 1994 and 1995 for vehicle speeds of up to 140 km/h. Furthermore, the smartcard could be used not only for tolling but for multiple payment purposes. Thus, the same 'personalized' card could be used for toll payments, buying fuel, paying for parking and as a travel card on public transport.

More direct pricing, with integrated payment using a 'cash-less' smart

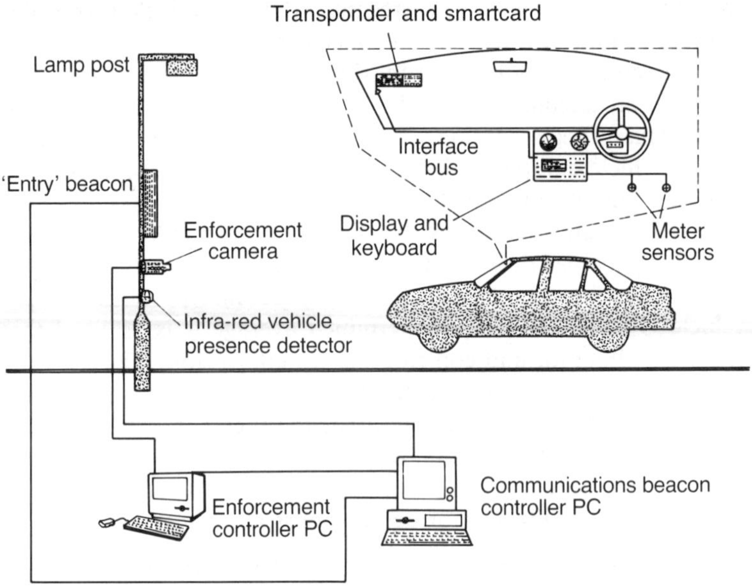

Figure 9.3 EC DRIVE II Programme - ADEPT project.

card system linked to on-line information services, offers a radical prospect for reforming the ways in which road space is used and charged for. However, the overall effect of these new developments will be merely to restrain the growth in traffic and improve its efficiency rather than diminish it. The impact on future estimates of fuel consumption and exhaust pollutants will therefore be slight.

9.5 ENERGY CONSUMPTION IN THE TRANSPORT SECTOR

Of all the sectors making up the total consumption of energy, transport now features as one of the larger ones. World-wide, about 20% of the 180 million barrels oil equivalent (boe/day) is down to the movement of goods and people. However, of the 75 million boe/day accounted for by crude oil (40% of all energy consumed), transport takes up 45%. Moreover, the dependence of transport on crude oil products is probably greater than any other sector. Also, as we have seen, it is set to grow faster than any other sector. Certainly, as the overall growth of primary energy supply shows (Figure 9.4), the huge increase since 1950 is very largely due to the growth in private car travel. At this scale, apparently major events (such as the OPEC 'crisis' and recurrent economic recessions) do not make any significant dents in the inexorable growth.

As far as the United Kingdom is concerned, the 42 Mtoe per annum of

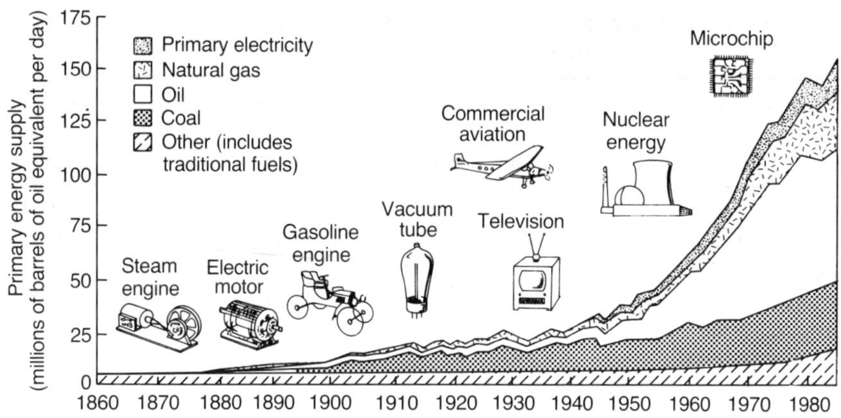

Figure 9.4 World energy consumption, 1860–1985.

petroleum products used by transport in 1993 accounted for just over half the national consumption. Road transport used 85% (35 Mtoe/annum) and this was split 70% petrol and 30% diesel. A small but highly significant element in this consumption pattern is the 6 million toe/annum which is consumed by aircraft, for which there is no conceivable alternative to fossil fuel. While it accounts for nearly one seventh of all the fuel consumed, it has increased by 35% over the last 10 years (despite very high prices for domestic air travel within Europe). The time may not be far ahead when the pollution consequences of air travel are regarded as more serious than those of car travel. This will depend, of course, on what progress can be made to curb the growth of fossil fuels consumed by land-based transport.

Ever since the internal combustion engine was invented just over 100 years ago, electric motors have offered them a recurrent challenge. Apart from tracked systems (railways and tramways), where the delivery problems can be overcome using mains overhead and conductor rails and storage is confined to serving ancillary power requirements such as the heating and lighting of vehicles, electric-propelled transport has been confined to a few small niches in the transport market.

That this should be so (and, probably, remain so) is due to some basic and essentially immutable physical laws, which govern the conversion and storage of electrical energy either chemically (via batteries and fuel cells) or mechanically (via flywheels and generators). As far as batteries are concerned, despite occasional reports of technical 'breakthroughs', they remain handicapped by three inherent limitations: (i) energy density; (ii) power density; and (iii) effective life. As Table 9.1 reveals, these limitations are very severe by comparison with the liquid fuel competitors.

Table 9.1 Comparison of specific energy and power outputs from different types of battery, as against liquid fuels

Type of battery	*Energy density (KWh/tonne)*	*Power density (KW/tonne)*	*Life (cycles)*
Lead/acid	20–40	50	400
'Improved' Lead/acid (30 °C)	50	66	800
Iron/nickel	60	55	1000+
Zinc/chloride	70	80	?
Zinc/air 'slurry' (55 °C)	140	160	2500
Sodium/sulphur (300 °C)	150	130	1500
Methanol	1200	200+	∞
Petrol (4 star)	2400	280 +	∞

The most 'advanced' prototype batteries, under laboratory conditions, can still only produce an order of magnitude less in energy density than a can of petrol.

The difference is less for power density but, for comparison on a like-for-like basis, the battery-driven vehicle would have a substantially lower proportion of its gross weight available for pay-loads. Moreover, the better the performance required in terms of acceleration, the bigger would be the draw-down on batteries; thus shortening their life.

Arguably, users might be persuaded to accept the severely reduced performance offered by battery-electric cars (for example, in urban areas as 'second' cars) but they would need to be assured of prospective gains to the community through improved efficiency and reduced pollution. Even if battery cars were introduced on a large scale, it is not clear that these gains would be achieved. For, despite the high efficiency of energy conversion in the electric vehicle itself, the generation and distribution of the electricity has a much lower efficiency. Indeed, when calculated in terms of overall thermal efficiency, battery cars show little if any advantage over internal combustion (i.c.) engines. Table 9.2 is based upon a broad set of assumptions concerning the relative efficiency of the different components of the two modes of propulsion.

The results suggest that, overall, the efficiency of both is in the region of 15%, comparing the useful work at the wheels with the primary energy input. In the case of battery propulsion, much depends on the efficiency of the electricity generation and distribution processes and how quickly it

Table 9.2 Comparison of relatively efficient battery electric vehicles vs. internal combustion vehicles

Output energy required by road vehicles (regardless of propulsion mode):

$$Z_0 = 100\text{–}150 \text{ Wh/tonne-km}$$

depending on performance required.

Relative efficiencies

Internal combustion			*Battery electric*		
Component	*Efficiency (%)*		*Component*	*Efficiency (%)*	
	1985	2005		1985	2005
Combustion	17	25	Battery	75	80
Drive-train	90	92	Control/drive	75	85
Oil-refining and transportation	90	90	Charger	85	90
(Syncrude)	–	(30)	Electricity gen./distrib.	29	33

Input energy required, therefore as a multiplier of output energy:

Mode of propulsion	*Efficiency (%)*		*Mode of propulsion*	*Efficiency (%)*	
	1985	2005		1985	2005
Internal combustion	$6.53Z_0$	$4.83Z_0$	Battery electric	$7.21Z_0$	$4.90Z_0$
(using Syncrude)	–	($14.49Z_0$)			

might be improved. Here again, battery vehicles may be disadvantaged because, using already-proved technology, a considerable reduction in the specific fuel consumption is likely with conventional i.c. engines.

In Figure 9.5, from a TRL study of the implications of the OPEC crisis reported by Porter and Fitchie (1977), a similar conclusion emerged, when comparing the total energy requirements in 2025 assuming all cars and (petrol-engined) light vehicles were converted to battery propulsion (or were reliant on Syncrude). Again, the battery cars offer no advantage in energy saving over i.c. engines, despite their higher cost and poorer performance.

One of these cost penalties which erodes the energy conservation potential of electric vehicles is the energy input required for battery manufacture. Another is the fact that electric heating of the interior of a battery car is needed in winter, since there is no 'waste' heat to utilize as

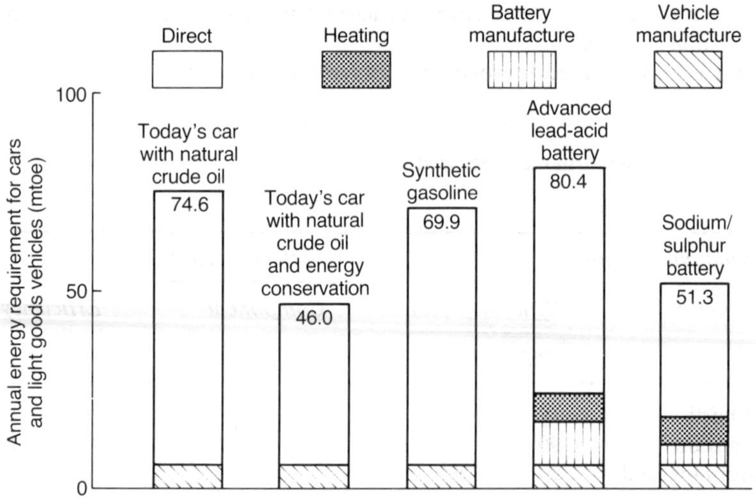

Figure 9.5 Annual energy requirements for cars and light vehicles in 2025. (Source: Porter and Fitchie, 1977.)

there is from a combustion engine. These two components combine to defeat even the sodium/sulphur battery's challenge to the best that 'lean-burn' gasoline engines could achieve. Although great hopes continue to be pinned on the development prospects for the sodium/sulphur battery, there are concerns over its safety in use. This is due to its 300 °C operating temperature, which could pose a risk if the cell were to rupture in a crash. (Similar concerns arise with flywheels, if they were to break free from their mountings following a collision). Of course, in each case, the strength of any outer casing would be designed to minimize these risks; but, in doing so, both weight and cost will be added.

9.6 STRATEGIES FOR REDUCING SPECIFIC FUEL CONSUMPTION OF CONVENTIONAL I.C. ENGINES

The total energy requirement calculated by Porter and Fitchie for 'today's' (i.e. 1977) cars in 2025 (74.6 Mtoe/annum) includes the energy input to its manufacture and assumes that saturation in car ownership would have been reached by then. It is clear now that saturation level is unlikely until 2030 or later; but, applying the NRTF high-growth figure (+ 142%) to the 1987 petrol consumption (24 Mtoe), gives a lower estimate of 55 Mtoe/annum – plus the 5.8 Mtoe/annum for manufacture equals 60.8 Mtoe/annum (compared to 74.6). This difference is explained partly by the higher car ownership figure they were assuming by 2025 and partly by the extent of fuel conservation already achieved over the 10 years following publication of their report. This conservation, of course, has been one of the beneficial consequences to flow from the OPEC crises.

The scope for further reductions in specific fuel consumption of conventional i.c. engines is considerable. A list of key strategies, all of which rely on known technologies, would include the following:

1. to introduce 'engine-management' systems, using microprocessor control of pressures, temperatures and air/fuel ratios;
2. to increase crank and drive-train efficiency, using new 'frictionless' bearings;
3. to change to diesel-powered or hybrid diesel/electric engines;
4. to limit engine size and move to a smaller capacity with fewer cylinders;
5. to recover waste heat from exhaust and cooling systems, using it to preheat injected fuel mixtures;
6. to reduce overall weight of vehicles, using aluminium or glass fibre instead of steel;
7. to smooth traffic flows and encourage steady driving, with fewer stop-starts.

With some or all of these improvements taken to their feasible limits, current fuel consumption rates for cars of (typically) 10–13 litres per 100 Km could be reduced to 6 litres per 100 km or less by 2025. This halving of specific fuel consumption would go a long way to offsetting the impact of more than doubling the amount of traffic. In fact, it would imply a rise of only 5 or so Mtoe/annum, from 25 Mtoe/annum in 1987 to 30 or so Mtoe/annum in 2025 on the high growth NRTF. It would also be substantially less than the 40 Mtoe/annum forecast by Porter and Fitchie for 2025. To the extent that new technologies for fuel conservation are available by then, as saturation in car ownership is approached, their figure of 40 Mtoe/annum of petrol in the United Kingdom may never be reached. Adding in the forecast consumption of diesel by heavy vehicles, and allowing for a similar achievement in specific fuel conservation, the total for road transport as a whole in 2025 should be in the range of 28–38 Mtoe/annum, corresponding to the low and high growth NRTF.

Relating this to the discussion of emerging transport policies (above), if the reform of vehicle taxation and introduction of road use pricing fail to be implemented, then the high growth NRTF is the more likely outcome. On the other hand, with a new system of charging for the use of road-space based upon marginal social costs of vehicle use, strong incentives to reduce specific fuel consumption, revenue support for public transport and widespread traffic restraint in urban areas, it should be possible to get total fuel consumption by road transport in the 2030s back down to that in 1990. Between now and then, however, under any scenario (apart from continual recession or war) total fuel consumption is bound to rise. This makes the achievement of the Rio Commitment of the UK Government on CO_2 levels impossible to achieve, at least as far as transport is concerned.

9.7 TECHNIQUES FOR REDUCING OTHER EXHAUST EMISSIONS BY ROAD VEHICLES

Apart from CO_2, the prognosis for cleaner exhausts and a steady reduction in exhaust emissions of all kinds is encouraging. The pressure to achieve these environmental gains, despite the prospective growth in traffic, comes from increasingly strict pollution control legislation which emanates mainly from California. Through the medium of a now global car market, what is good for California will (it seems) be good enough for us all.

Among the more promising, techniques for reducing specific exhaust emissions by road vehicles are the following:

1. to introduce 'engine-management systems' (as with fuel conservation);
2. to retard the ignition and optimize valve operations to provide the requisite power;
3. to recirculate exhaust gases, up to 10% by volume (as with fuel conservation);
4. to fit three-way catalytic converters to all new cars;
5. to change to fuel injection, with 'designer' additives to the air/fuel mixture to boost octane value;
6. to adopt 'lean-burn' technology, based upon 'stratified charge' of air/fuel mixtures, to provide requisite power on demand;
7. to move to new forms of ignition (e.g. plasma spark or catalytic) to reduce the time and increase the efficiency of combustion.

With some or all of these measures, taken to their feasible limits, most exhaust pollutants could be significantly reduced (with the exception of CO_2, which is likely to be augmented!).

Although the introduction of the catalytic converter to meet the most recent pollution-control standards will have a marked effect on most pollution levels (especially if they can be made more effective at low exhaust temperatures), the adoption of lean-burn technology using engine management systems will be the more significant step. Not only is this a move away from 'end of pipe' treatments, which inevitably carry a fuel consumption penalty, but it would usher in an era of engine design with the twin objectives of minimizing both fuel consumption and exhaust emissions. With this synergy established, substantial gains can be realized. Nevertheless, the advantage of the interim step of fitting three-way catalytic converters can be seen, at least as far as 'greenhouse' gases are concerned.

Thinking of implications for the global environment, however, the determinants may well lie beyond what any one country (or group of countries) is able to achieve. In this context, the prognosis is less encouraging.

9.8 GLOBAL IMPLICATIONS OF FUTURE GROWTH IN TRAVEL DEMAND

If we look at the vehicle ownership growth in a world perspective, we can see that the population of vehicles has grown 12-fold since 1945, from 50 million or so then to about 600 million now. Within this total, the distribution bears little relationship to that of population. One third of the vehicles are registered in the United States, which has only 6% of the world's population; one third are to be found in Western Europe and Japan, which together account for 10% of population; and one third of the vehicles are owned by the remaining 84% of the world's population.

As saturation in car ownership is achieved in the already wealthy car-owning nations, these proportions will change, with the vast preponderance of growth in vehicle population taking place in the middle-income countries. It will be what is done to limit the growth in these countries, rather than the already wealthy ones, that will determine future demands for fossil fuel and future impacts on the global environment. For all our sakes, we should hope that they will find ways of living in future that depend rather less on car use that has been the case with us. If not, then the prospects for keeping global impacts in the environment within manageable limits seem rather remote.

Two conflicting pieces of evidence indicate the difficult struggle ahead. One, an optimistic indicator, reveals that technological transfer is highly effective; such that 'best practice' and high standards of efficiency are passed on down the chain of economic development. Each successive country to undergo the industrializing process does so at a lower level of energy intensity than its predecessor. Thus, for each US$1000 of GDP added, the equivalent amount of energy required (in Mtoe) falls, once the 'threshold of development' is passed. This suggests that the energy implication of any given level of economic development is steadily reducing. In the case of transport, this could be exemplified by the progressive replacement of demand for physical movements of people by telecommunications, with orders of magnitude savings in energy cost. Developments in transport 'telematics' offer considerable promise in this respect.

The pessimistic indicator is that revealed by the failure to comprehend the importance of reforming the taxation and pricing structures of transport. As Figure 9.6 shows, there are wide discrepancies between the retail prices charged in different countries for a gallon of gasoline. As a world-wide traded commodity, the resource cost of production and distribution (60–80 US¢ per gallon) varies only slightly between different regions. The huge differences in price are due to governmental taxes and

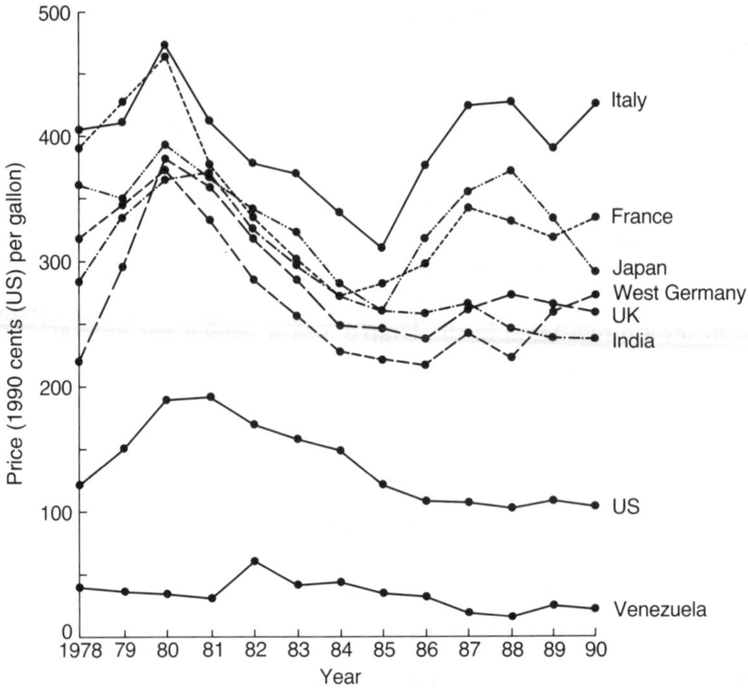

Figure 9.6 Retail prices of gasoline in various countries, 1978–1990. (Source: Amoco.)

excise duties. Alone among the developed, oil-consuming countries, the United States imposes only nominal taxes on petroleum products and yet its population consumes one third of the world's oil production. Of the two developing nations shown in Figure 9.6, India as an oil importer (despite its very low income per head) taxes gasoline at almost the same rate as the United Kingdom and Germany. Venezuela, on the other hand, as an oil exporter (with a moderate income per head) hardly taxes the product at all. Indeed, the Venezuelan price scarcely covers even the local costs of production.

With the example set by the United States and Venezuela, how can we hope that the countries of the developing world, as their demands for vehicle use rise, will act to restrain this growth by prudent fiscal policies and effective price mechanisms? In the absence of this restraint, the future demands for fossil fuels will be enormously greater than they are now and the implications for the global environment correspondingly more serious. By the time China, Indonesia and the Philippines (for example) have all expressed their demands for private car ownership and use, our current efforts at pollution control may pale into insignificance. Alternatively, we will succeed in transferring the best technological practices to these countries, so as to minimize the environmental consequences of their wholesale motorization.

REFERENCE

Porter, J. and Fitchie, J.W. (1977) Energy for Road Transport in the UK, Supplementary Report SR311, Department of Transport (TRL), Crowthorne.

FURTHER READING

BP (1989) *Statistical Review of World Energy: July 1989*. British Petroleum plc, London.

Department of Transport (1989) *National Road Traffic Forecasts (Great Britain) 1989*. HMSO, London.

Department of Transport (1993) *Transport Statistics: (Great Britain) 1991 and 1992*. HMSO, London.

Department of Transport (1993) Paying for Better Motorways (Green Paper discussion document). HMSO, London.

Energy for Planet Earth *Scientific American* (Special Issue), September 1990.

Hills, P.J. and Blythe, P.T. (1993) Annual Review Report on ADEPT project, submitted to CEC DRIVE Central Office, Brussels, October.

Institution of Highways and Transportation/Department of Transport (1987) *Roads and Traffic in Urban Areas*. HMSO, London.

Institution of Highways and Transportation (1993) *Transport and the Environment*. Position Paper No. 3, IHT, London.

International Road Federation (1992) *World Road Statistics (1988–92)*. IRF, Geneva and Washington.

RAC Foundation Advisory Group (1993) *Cars and the Environment: a View to the Year 2020*. First report to the RAC Foundation, London.

Waters, M.H.L. (1992) *Road Vehicle Fuel Economy*. TRL State-of-the-Art Review No. 3, HMSO, London.

Watkins, L.H. (1991) *Air Pollution from Road Vehicles*, TRL State-of-the-Art Review No. 1, HMSO, London.

10

Energy efficiency and approaches to it

E.W. Lees

10.1 INTRODUCTION

Countless studies of the best ways to satisfy the energy services required either on a local, regional or global scale invariably come to a conclusion that energy efficiency and energy conservation should play a much greater part in meeting these service requirements than they currently do. However, we all know the reality of the situation – that energy saving measures are not taken up to the extent that all these theoretical studies predict.

This chapter will address the issues of why energy efficiency and energy conservation are important, why the technology is not exploited to the full, the policy framework within which exploitation has to occur and in particular the structure of the utilities and why the author believes there is still significant scope for new technologies to make significant energy saving contributions in the third millennium. Finally, the author turns to one of his hobby horses, i.e. the necessity of energy efficient technology being considered within a well-managed environment.

10.2 THE IMPORTANCE OF ENERGY EFFICIENCY

The use of energy has become essential to all aspects of modern life and economic activity. The United Kingdom alone consumes nearly 2600 TWh of energy per year, equivalent to over 6 tonnes of coal for every man, woman and child in the country. Yet the benefits of energy come at a price, both economic and environmental. The United Kingdom spends over £50 billion a year on energy. As well as contributing to the 'greenhouse' effect, burning fossil fuels also causes air pollution in the form of dangerous oxides of nitrogen and sulphur, as well as smoke particles and hydrocarbons. Nuclear power generation has its own problems with radioactive waste and contamination risks. Even renewable sources of energy are not without environmental impacts.

Energy conservation and energy efficiency activities have been undertaken in the United Kingdom and world-wide on the grounds of:

1. energy security;
2. preserving resources of fossil fuel for future generations;
3. economic efficiency;
4. avoiding the construction of new power plants;
5. reducing environmental pollution;
6. minimizing damage to health.

Energy saving brings all these benefits and energy efficiency must therefore be central to any sustainable development strategy. In the United Kingdom, the emphasis is currently on the environmental benefits which energy saving can bring but the other benefits are always present and will become increasingly important as concerns about health in urban environments and the concept of sustainable development are taken further. In the final analysis, energy efficiency is a way of continuing to enjoy the same or improved levels of energy services, while reducing the actual energy required to provide them. This saves money and helps the environment.

10.3 WHY IS ENERGY SAVING NOT EXPLOITED TO THE FULL?

There is significant energy saving potential available in the United Kingdom today – the EEO's estimate is that at least 20% or £10 billion per annum could be saved; however, there are a few barriers to the exploitation of this energy saving potential.

Much has been written about barriers to energy efficiency which are divided into five main classes:

1. information barriers;
2. economic barriers;
3. hidden costs/'hassle' barriers;
4. market imperfections barriers;
5. public acceptability barriers.

Let us consider each of these in turn. Firstly the familiar **information barriers** are the lack of awareness and reliable information on the part of both business people and householders of the opportunities to reduce energy consumption particularly at the point of decision making.

Perhaps the most important **economic barriers** are the much higher rate of return required by businesses and householders for investment in energy efficiency and the lack of capital (particularly important for low income householders). In reality, these barriers reflect the 'peripheral' manner in which both businesses and householders view energy efficiency.

Examples of **hidden cost barriers** include the cost and time involved in acquiring information on energy efficiency measures and even the cost of implementing these. Additionally, the implementation of an energy efficiency measure may occasionally have risks associated with it (particularly true for novel measures) and acts as a further disincentive.

There are many **market imperfections** but perhaps the most notable are price distortions. For example, in the United Kingdom the gas tariffs in recent years have not encouraged energy efficiency and certainly the current prices of fossil fuels do not reflect the costs of environmental damage associated with their production and consumption. The **'landlord/tenant' barrier** is the general name given to the situation in which equipment (such as heating systems in rented accommodation or motors in much electrically driven machinery) are specified and purchased by people who do not have to meet the subsequent running costs. This barrier would also include institutional constraints, e.g. public sector finances frequently have capital restrictions placed upon them.

The final barrier of **public acceptability** is one which is frequently overlooked. For example it is clear from comments made by consumers that there is, for some, a genuine resistance to the introduction of energy efficient light bulbs. Most of the complaints really refer to the loss of amenity compared to the performance and light colour of the traditional filament lamp, and recent improvements in the design and manufacture of these lamps have gone a long way to alleviating these objections.

The UK government policy in recent years has concentrated too much on the information barriers. Some of the market imperfections have been tackled by the UK Government throughout regulation, e.g. new buildings, appliance labelling, but with the notable exception of the Home Energy Efficiency Scheme which subsidizes low income households where the investment barriers are greatest, very little has been done to tackle the economic and hidden cost barriers.

This is one of the reasons why the author accepted the challenge of being the founding Chief Executive of the Energy Saving Trust. By levying gas and electricity consumers, the funds raised can be given back over time to those same consumers to encourage them to invest in energy efficiency and energy conservation measures which are already in their own economic interest. It was recognized that the Trust had the opportunity to tackle such barriers in a fashion which has benefits for the individual consumer, for the nation as a whole and for the environment as a result of reducing energy demand.

10.4 POLICY/UTILITY STRUCTURE ISSUES

The implementation of measures to save energy in the regulated utilities (gas and electricity) which provide 72% of our non-transport energy requirements, is clearly paramount. In turn, this depends crucially on government policy and the structure of the utilities. When the UK Government signed the Framework for Climate Change Convention in Rio in 1992, it clearly recognized that a significant proportion of the required reduction in CO_2 savings would have to be delivered through the actions of the major power utilities. The question which remained unanswered was whether the UK electricity industry - with its competitive and highly disaggregated structure - would be able and willing to respond to the challenge of instigating large-scale demand side management (DSM) programmes. The electricity industry is used as the UK example as its disaggregated structure is further advanced than that of the gas industry. However, comments made are likely to be applicable to gas in the future.

In the space of a little over 5 years, the United Kingdom has moved from the possession of a traditional, vertically integrated monopolistic, nationalized electricity industry to to have one of the most disaggregated and competitive utility industries in the world. By competition, this means not only competition in generation, but also, almost uniquely, significant competition in electricity supply though not for small users, i.e. domestic customers and small businesses.

Before privatization, generation and transmission in England and Wales were dominated by the Central Electricity Generating Board (Figure 10.1). After privatization, this was split into three generating companies and a separate transmission company. These generation companies plus the independents can sell into the pool and direct to large customers (Figure 10.2). The National Grid Company controls the transmission system and ranks the generators' price bids on a daily basis and is responsible for the dispatch of electricity. The so-called non-franchise customers (greater than 1 MW currently and greater than 100 kW from April 1994) can purchase electricity from any regional electricity company or any other licensed supplier or direct from the pool. One-third of non-franchise

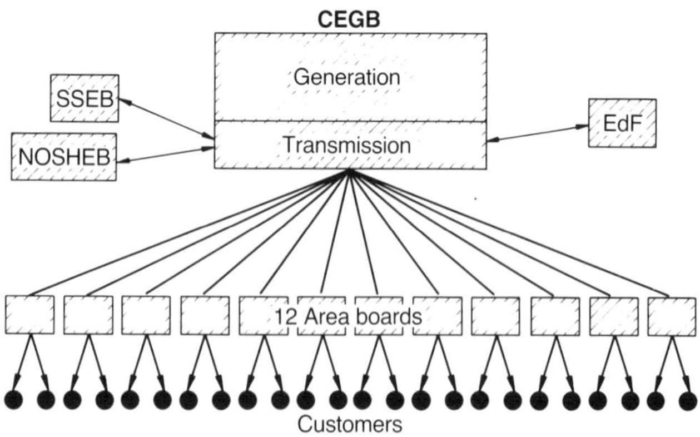

Figure 10.1 The UK electricity industry before restructuring.

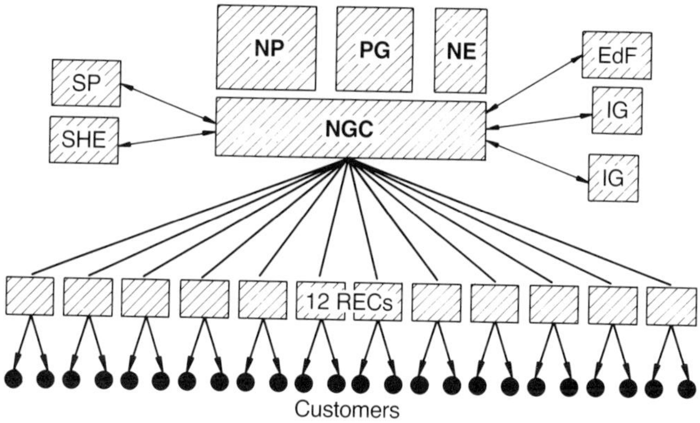

Figure 10.2 The UK electricity industry after restructuring.

customers have chosen a supplier other than their local regional electricity company.

The role of the Electricity Regulator within this new structure is to promote competition, oversee prices, protect customer interest where a natural monopoly remains, prevent cross-business subsidy and to consider the effect on the physical environment of the electricity companies' activities. For the franchise customer perhaps the two most obvious aspects of the Regulator's activities are in the control of prices via the famous RPI – X + Y and the area of standards of performance, including security and safety of supply, customer care/service and energy efficiency.

The Regulator has decided to allow a £25 million per annum allowance under the supply price controls while at the same time removing the majority of the volume incentive within the old price formula and hence encouraging DSM as an alternative to increased supply. The distribution price control is currently under review and the regulator has asked for comments on volume effects, distribution losses and financial incentives for utilities to undertake DSM.

People often try to take the US experience and translate it to the UK situation. It is very important that we do learn as many lessons as possible from the United States but equally we have to bear in mind the differences between the two countries not least in their regulatory approaches. In the United States, where significant DSM schemes have been implemented, these were originally driven by restrictions on capacity and, in particular, local opposition to building new plant. There is an emphasis on least cost planning and, in general, the US regulators are working with vertically integrated utilities where there has been a long history of strong regulatory input. By contrast, in the United Kingdom it is to a certain extent environmentally driven, justified on consumer benefits and addresses a disaggregated industry where there has been a recent, limited regulatory input. Therefore it is necessary to adapt the US experience to the UK situation.

Nevertheless it remains the conviction of the Trust that energy efficiency measures are possible even within the disaggregated utility industry but that the role of the Regulator and his or her statutory powers need to be clearly defined. There is no doubt that there needs to be a strong institutional focus to ensure that these measures are implemented to the full and the author believes that this is an important role for the Trust to fulfil. If significant energy savings are earnestly sought, this will inevitably mean significant expenditure and what is more real reductions in the load of the utilities. This can only be successfully achieved with the full support of the utilities and argues for incentives for utilities to cut their load at least equal to those for increasing their load. All in all, it is an attempt to help them move from fuel utilities to energy service companies.

10.5 ROLE OF NEW TECHNOLOGY

The author does not want to give the impression that nothing is happening in energy efficiency – simply that it is not happening either fast enough or to the extent which is environmentally and economically justified from a national perspective. For example, the efficiency with which energy is used in manufacturing industry has increased dramatically throughout the United Kingdom and indeed all OECD countries since the first oil crisis. If we take the example of annual industrial energy use, this peaked at 2720 PJ in 1973, since when it fell to 1623 PJ in 1990. Over the same period, the value of industrial production in real terms has changed very little. So the output per unit of energy of the UK manufacturing sector has increased by approximately 40% over this 17-year period (Figure 10.3).

In the years following the two oil crises, the increased price of energy acted as a spur to management to pursue energy efficiency and this was reflected in a lowering of energy intensity. But it is not a simple relationship. Figure 10.3 shows that since 1985 real energy prices have fallen, so that now they are at much the same level as in 1973. Yet, at least in manufacturing industry, the trend towards higher energy productivity (lower energy intensity) continued unabated until the onset of the 1990–93 recession.

Of course, these aggregated trends mark at least four different forces at work.

1. Structural changes, i.e. growth of less energy intensive industries, e.g. the rise of non-intensive sectors such as electronics, while older energy intensive industries such as steel have declined.
2. At the sectoral level there has also been a move towards higher quality, higher added value products at the expense of bulk products.

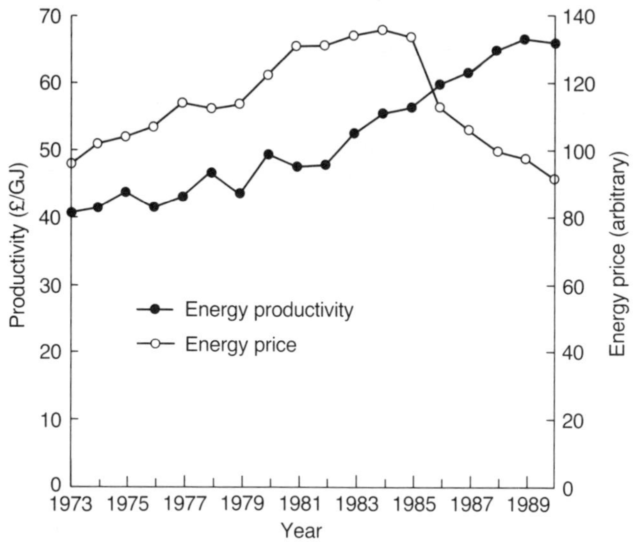

Figure 10.3 Energy productivity in UK manufacturing.

3. Increased management awareness that energy is a controllable resource through better control systems and improved management technique.
4. Technologies and techniques designed to improve the efficiency of energy use have been developed, demonstrated and marketed, particularly in the energy intensive sectors.

In thinking about the role of new technology in improving energy efficiency, it is common to concentrate on the last of these categories. Yet this is a mistake. Growing industrial sectors and new products are the main targets for industrial investment and provide important opportunities for energy efficient innovation. Very often, energy efficiency improvements flow from projects designed to introduce a new product, to increase labour efficiency or to improve product quality. For example, the requirement for consistent quality in aluminium castings for high performance petrol engines provided the stimulus for Cosworth to develop new diecasting technology. The resulting Cosworth process not only gave a high quality product but also more energy efficient foundry practice.

New technology has also been a key part of the changes in monitoring, control and management information. The expert system package, Linkman, developed by Blue Circle for more accurate and consistent control of rotary cement kilns, has given improvements in energy efficiency, typically in excess of 5% of energy consumption in a very energy intensive process. Similarly, in the building sector, building energy management systems (BEMS) became commonplace in large commercial buildings during the 1980s. In both cases energy efficiency benefited from the rapid improvements in low cost data processing which flowed from the availability of microtechnology.

The key to increased energy efficiency is unlikely to be largely through a single technological breakthrough but to ensure that energy is fully considered in all management decisions – from the frequency of boiler servicing to the design of a new factory. New technology needs to be considered in this context.

10.6 POTENTIAL SAVINGS FROM NEW TECHNOLOGY

New technologies can be divided into two broad categories: those which are available at present but are not widely used, and those which require further R&D before they can be successfully deployed. The energy savings potential from the latter category is obviously very uncertain. Even for the former category there are considerable problems in estimating the potential for energy savings, as the following examples taken from ETSU's studies on industrial energy efficiency illustrate.

The reduction in energy use which would occur, if UK industry adopted all process energy efficiency opportunities ETSU has identified which offer a return on capital of 8% or greater, would be 23%. As these technologies are already cost effective, these savings could technically be achieved within this decade without any problem. A fall in energy use of this magnitude would give an increase in the rate of energy productivity improvement from historic levels shown earlier without any allowance for continuation in the trends in structural change.

The identification of this level of untapped energy efficiency is far from being a unique or uncorroborated conclusion. Other analyses have shown similar levels of cost-effective opportunities and there is some evidence that in the domestic and transport sectors the theoretical untapped potential is higher than in industry. For example, a Commission of the European Communities project, entitled Cost Effectiveness Analysis of CO_2 Reduction Options, studied the role of various technologies in achieving abatement of carbon dioxide emissions. Figure 10.4 shows that for the Community as a whole, energy conservation measures constitute the largest component of the reductions. More importantly, it was also concluded that the vast majority of the energy efficiency measures identified were already cost effective. In view of the short payback periods required, energy efficiency is clearly 'economic good sense' as well as environmentally beneficial.

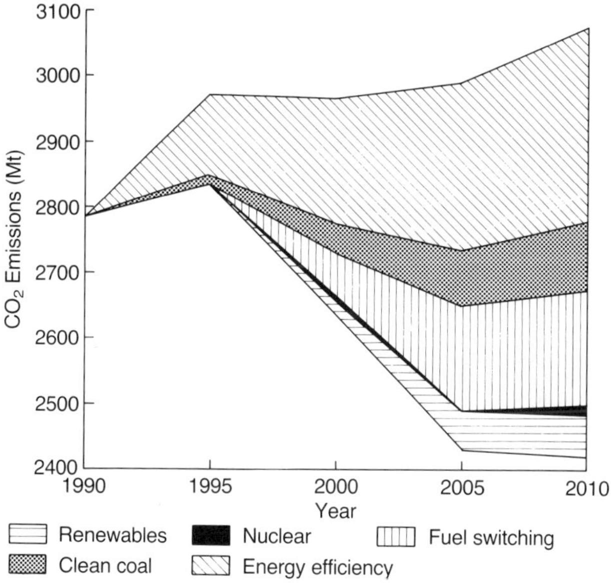

Figure 10.4 Contribution of technology to abatement of CO_2 in the European Community.

10.7 REALISTIC PROSPECTS FOR INDUSTRIAL ENERGY EFFICIENCY

The explanation for these apparent inefficiencies is partly in the investment criteria used. We know that most companies do not invest on the basis of the 8% rate of return typical of government. Surveys of the manufacturing sector indicate that 25% is claimed to be a typical criterion for investments in the category of discretionary cost saving projects, under which most energy efficiency measures fall. The same analysis for UK manufacturing industry is undertaken using a 25% discount rate. Of course the level of energy investment is lower and the potential savings are smaller. But there is still the potential for an 18% reduction in energy use without any allowance for structural change in the manufacturing sector.

The small difference between the energy saving potentials for the two discount rates (8% and 25%) is rather surprising and reflects the fact that most of the identified energy efficiency opportunities are cost effective at both discount rates. In other words, technologists reflect the strict investment criteria of the industrial market place in their selection of R&D projects and only investigate the most attractive energy efficiency opportunities. The corollary of this is that it is likely that there are many more opportunities, which have not been identified or quantified, that would prove cost effective at an 8% discount rate. Although it is interesting to speculate on what the size of the potential saving opportunities might be in an idealized 8% discount rate world, in the real world the current stringent criteria for energy efficiency investment mean that most opportunities are unlikely to be investigated in much detail.

Even the most ambitious proponents of energy efficiency would not expect all this 18% potential energy saving to be realized very rapidly.

Of course, similar factors affect most investment opportunities, but they are particularly relevant to energy efficiency. Energy is peripheral to the major concerns of most industrial decision makers, and therefore is relatively easy to neglect. The result is that our database and analysis, which assume economic optimality, describe what might happen in an ideal world, not unfortunately what will happen in the real one. The lessons of this and similar analyses, and the questions which flow from them, have been summed up by Michael Grubb (1990) in his book *Energy Policies and the Greenhouse Effect*.

> Presuming that, given time, the market alone can deliver optimal energy efficiency is therefore facile. Projections which assume that the full theoretical potential for efficiency can be captured are equally so. But the economic and environmental potential stands beckoning. What, then, are the policy options for trying to narrow the 'efficiency gap', what impacts might they have and how might Governments choose among them?

Of course, there is no easy single answer to this question. As Helga Steeg of the IEA has clearly stated, 'Energy is not the classical example of a freely-traded good in a market economy' (15th World Energy Congress, Madrid 1992). The IEA has recommended that their member government should agree compatible policies which 'support the efficient functioning of healthy and transparent energy markets'. The IEA believes that there is no single prototype energy policy in market economies and that policies need to reflect the capacity of individual countries.

10.8 THE WAY FORWARD – TECHNOLOGY AND MANAGEMENT TOGETHER

Turning to the industrial sector, policies need to be geared both to promote the exploitation of new technology which already exists and to ensure that R&D continues to provide further generations of new technologies. Which policies can contribute to this? Exhortation, taxation and legislation all undoubtedly have a part to play. But our experience of 'selling' energy efficiency to the UK private sector tells us that these should not be seen as isolated initiatives.

The key to the effective use of new technology for improved energy efficiency is good management. What is needed is a management culture which recognizes that the commercial pressures to reduce energy use are no longer the result of a short-term oil crisis and that environmental concerns are more than a 'temporary green craze' They have to be recognized as part of a long-term shift in the realities of the economic and environmental landscape. Only this can generate widespread management attitudes which treat energy as an important resource and recognize that both its financial cost and its potential damage to the environment are controllable. The following is a brief description of a UK example where such a shift in management attitudes has led to significant energy savings.

The approach is to try and tie energy efficiency considerations into the total quality management or quality circles which are becoming increasingly widespread throughout industrialized countries. For instance, H.J. Heinz has a world-wide mission, to become 'the most energy efficient food processor using the skills of our employees and outside bodies'. Hence in the United Kingdom the energy manager has the full support of the UK managing director in his campaign to pursue greater energy efficiency. The fact that the message starts at the top is a very important point and is one of the features which is most clearly identifiable from successful case studies.

However, the message has to extend all the way through the company and Heinz has achieved this by extending the Heinz total quality management systems to include energy management. Each profit centre is accountable for its energy use, which of course extends ownership of energy to the end users. Employees have been encouraged to think of, and to implement, their own energy saving ideas. Heinz operates a profit sharing scheme for its employees and so the connection between energy saving and profit is being stressed to provide an incentive. Finally, an incentive was established by allowing 10% of the savings to be given to a local charity of the quality circle's choice. The campaign saved £230 000 without expenditure on energy saving equipment (over 5% of previous consumption).

Technologies by themselves are not sufficient; they have to be managed and integrated into the normal resource management activities of a company. Without effective management, even the best technologies for energy efficiency and fuel switching will have only limited results. This more businesslike approach to energy management will be of benefit, not just in terms of energy saved, but also will make the adoption of innovative energy saving technologies or techniques easier in the future.

REFERENCE

Grubb, M. (1990) *Energy Policies and the Greenhouse Effect.*

11

UK energy strategy

I. Fells and J. Horlock

11.1 INTRODUCTION

The United Kingdom is in the fortunate position of having indigenous supplies of coal, oil and gas, as well as a nuclear component of electricity generation. This has led to a degree of complacency; shortages, it has been believed, are unlikely.

Against this background, the Government stated that 'competitive markets provide the best means of ensuring that the nation has access to secure, diverse and sustainable supplies of energy in the forms that people and businesses want, and at competitive prices'. The Government's energy policy has therefore centred on the creation of competitive markets, and this has recently led to a major change in the balance of fuel use (the 'dash for gas' and the substantial reduction in electricity generation from coal-fired power stations).

However, the market pays little heed to the environment, which it has until recently regarded as 'free'. This situation is changing as the European Community develops its energy policy through legislation and regulation, and as stabilization of 'greenhouse' gas emissions becomes an international imperative.

Matching technological developments to an environmentally constrained energy policy is a difficult task. Engineers are good at solving technical problems but many aspects of energy use are dominated by economics and social attitudes. Options for the long-term future are reviewed in the chapter; a way forward for the United Kingdom is proposed, as an alternative to the present policy, which is solely market-led.

The United Kingdom is an important member of the European Community and has signed the Energy Charter. UK energy policy must therefore fit into a European energy policy and is already constrained by environmental legislation such as the large combustion plant directive which deals with the minimization of acid emissions. The proposed Energy

Charter carbon tax could have a significant effect on future UK fuel mix, with coal seriously disadvantaged. Emissions from road transport are also controlled with ever tightening legislation. Stabilization of 'greenhouse' gas emissions is an international imperative, viz. the Rio Conference and Protocol of 1992. It follows that over the next 10–20 years, energy policy will be strongly environmentally influenced as well as resource-led.

The United Kingdom's position regarding supplies of coal, oil and gas has led to the theory that should shortages become evident then they can be accommodated by switching fuels, with plentiful supplies of coal always a fallback position. The 'dash for gas' is the current enthusiasm and it is argued that possible shortages can always be supplemented by imports from Russia, Norway or the Middle East. (What will happen to the future price of gas is another matter.)

However, the Government White Paper *The Prospects for Coal* (Cm 2235, HMSO, 1993) clearly spells out the UK energy policy as follows. '... Competitive markets provide the best means of ensuring that the nation has access to secure, diverse and sustainable supplies of energy in the forms that people and businesses want, and at competitive prices. The Government's energy policy therefore centres on the creation of competitive markets.'

11.2 RECENT HISTORY LEADING TO THE PRESENT POSITION

The effect of this policy on electricity generation has been dramatic over the past five years. Two key events – the freeing of restriction on the use of gas and privatization of the electricity supply industry – have produced substantial changes in the pattern of generation. Table 11.1, taken from the National Grid Company's 1993 seven year plan, shows NGC's figures for generation capacity (the sum of 'transmission contracted' and existing) and the various fuels used. Coal is held at a level of about 30 GW, nuclear increases with the introduction of Sizewell B in 1994/1995, but the major feature is the substantial increase in the amount of electricity to be generated by the new, efficient combined cycle gas turbines. Some 23 GW of CCGT plant – getting on for half the electricity demand in the United Kingdom – may be introduced. Table 11.1 therefore implies a developing surplus of generation capacity over demand, but note that not all the 'transmission contracted' plants will necessarily be built and some of the existing plants will no doubt be closed.

The reasons for this enormous projected growth in CCGT plant are its high thermal efficiency, its low maintenance cost, its low capital cost and the speed with which it can be built. These factors lead to low pricing of 'units sent out'.

Fuel costs form a high percentage of the levelized costs of CCGT plant (some 50–60%), the capital costs being relatively low. Thus for CCGT

Table 11.1 UK generating plant capacities (MW)

	1993/94	*1994/95*	*1995/96*	*1996/97*	*1997/98*	*1998/99*	*1999/00*
Nuclear	9 524	10 873	10 933	10 933	10 933	10 933	10 933
Coal	30 271	30 271	30 271	30 271	30 271	30 271	30 271
Oil & OCGT	10 634	10 634	10 634	10 634	10 634	10 634	10 634
CCGT	4 875	9 049	13 514	18 335	20 412	21 612	23 047
Other	4 922	4 922	5 672	5 672	5 672	5 672	5 672
Total	60 226	65 749	71 024	75 845	77 922	79 122	80 557

plant there is considerable pressure to hold gas costs, and to increase thermal efficiency even further. For nuclear plant, as far as levelized costs are concerned, the fuel costs are a small fraction, capital costs forming some 70% of the total. The emphasis, therefore, is to reduce the capital cost of new nuclear plant, together with the costs of decommissioning and waste disposal.

The avoidable costs of existing nuclear plant (with capital written off) are very low, so there are strong arguments for keeping them running as long as the costs of maintaining high safety levels do not become uneconomic. As far as coal, is concerned, avoidable costs may fall below CCGT levelized costs; but other factors are the cost of introducing fuel gas desulphurization, the price of the coal itself and the fact that the bigger plants are generally of higher thermal efficiency than the older, smaller, plants.

All these factors were taken into account in the Government's coal review, that review being essentially based on market forces rather than a long-term strategy, including environmental factors. However, it must be said that the new CCGT plants are environmentally friendly, with the natural gas producing low sulphur in the flue gases and lower CO_2 production per unit sent out (kg CO_2/kWh). Here concern for the environment is fortunately coinciding with the drive of market forces.

11.3 INFLUENCES ON FUTURE POLICY

11.3.1 Influences within the United Kingdom

More usually the market pays little heed to the environment, which it regards as free. In consequence the environment can only be protected by regulation, via taxation, limits on emissions, etc.

The United Kingdom has probably gone further than any other country in 'freeing up' its energy industry; only coal and nuclear power remain under government control, and the introduction of measures to increase competition are given priority.

Nevertheless, there are elements of policy which distort the energy market. Non-fossil fuel is subsidized, with some generators receiving two or three times as much per unit as gas- and coal-based generators. Substantial grants are now becoming available in the formerly neglected energy efficiency area, for the purchase of more efficient boilers for example.

It has to be said that these aspects of energy policy have been applied in a somewhat piecemeal manner. The energy demand side is policed by regulators providing surrogate competition, where suppliers are in a monopoly or oligopoly position. UK energy policy is in reality, therefore, a construct, which is constantly being tinkered with and not the rather grand, simplistic, market-led policy which the Government advertises.

The long lead times in energy planning and development point to the main danger inherent in the market-led policy (however modified) – simply that short-term effects dominate. The recent moves to gas-fired electricity generation in the United Kingdom provide a potent illustration of a market-led action, without regard to the effect on the indigenous coal industry, or for that matter the nuclear industry. Questions about security of supply and social costs do not enter into the equation. The future price of natural gas, *vis-à-vis* coal, is seen as satisfactorily accommodated by 15-year contracts with appropriate escalation clauses. However, power plants are built to last 30 to 40 years.

11.3.2 European influences

An important ingredient of a competitive European energy policy, which signatories of the European Energy Charter will gradually be encouraged to introduce, is third party access; that is the, free movement of energy across the national boundaries of the European Community. This applies particularly to gas and electricity where the distribution system is a natural monopoly. Supplies of nuclear electricity are already traded from France into Austria, Spain and the United Kingdom and from Sweden to Denmark. Natural gas is imported from Russia and Norway. Natural gas provides 19% of the total energy supply to the Community, of which about 50% is imported.

If a truly competitive system is to arise, new energy suppliers must have access to gas and electricity from countries of origin, and the ability and permission to export across national boundaries at reasonable rates. Some form of regulation will be necessary and the European Energy Charter has been designed to provide the necessary framework. Progress will be slow; it is anticipated that the complete freeing of the European energy market may take as long as 20 years. Here UK experience in developing third party access will be valuable.

11.3.3 World influences

But the changing demand for energy on the international scene will have a profound effect on future supplies of energy to the European Community. World energy demand is expected to increase by between 50% and 70% by the year 2020, over current values (World Energy Council); even with a strong ecologically driven programme the increase will be at least 35%. This is partly accounted for by an anticipated doubling of world population between 1990 and 2050, when developing countries will account for over 85% of the total number. The enhanced expectations and rapidly increasing economies of many of the developing countries, particularly around the Pacific Rim, account for this inexorable demand, even taking into account considerable improvements in the efficiency of energy use. An increase of between three- and fourfold over 1990 levels of demand are anticipated by the year 2100.

11.3.4 Future energy policies

Against all these influences and background, a plan which develops a balanced energy portfolio for the United Kingdom over the next 50 years or so may be dismissed as a Utopian dream discredited by the failure of the Central Planning Procedure of the former Soviet Union and its satellites. This is to ignore the enormous strides made in modelling and scenario planning procedures in the energy and environment fields, and also to ignore such conspicuous planning successes as the French nuclear programme. Before proposing such a strategic energy policy, we will consider the technological situation.

11.4 TECHNICAL DEVELOPMENTS IN ENERGY USE

Many technological advances are available, or are capable of development, to reduce energy use, both in efficient utilization of primary energy and in reducing demand.

In electricity production, CCGT plant efficiency is still rising with higher turbine temperatures, use of reheat and triple pressure steam raising. The latest CCGT plant offered by ABB has a remarkable thermal efficiency of 58.5%.

On coal-fired generation, developments in clean coal technology are leading to reduction of SO_2 production, but only increased thermal efficiency will reduce the CO_2 per unit of electricity sent out. The best prospect for long-term utilization of coal would appear to lie in its use in combined heat and power (CHP) plant, with high overall plant effectiveness (utilization of both heat and electrical work output as a fraction of primary energy input); the CO_2 is still inevitably produced from the coal but its primary energy is better used than in a simple coal power plant.

Nuclear technical developments are likely to involve reduction of the capital costs (reducing the time of construction, for example, and building a series of identical plants) and improving safety still further (the introduction of plants with passive safety features built in) and reducing the cost of decommissioning and waste disposal.

The prospects for renewables must depend on some degree of subsidy. Most renewable proposals, although within range technically, look uneconomic unless the discount rate is low or subsidy is introduced.

As far as the other major sectors of energy utilization are concerned - industry, domestic heating and transport - developments in and introduction of high energy efficiency are crucial for the first two. Small CHP has done well in industry since the Marshall report, but large CHP involving district heating has not been introduced in the United Kingdom - it requires low discount rates to be economic. So improved insulation and increased boiler efficiency must be the front runners in the domestic sector, rather than the more exotic solutions of solar heating and electrical lighting equipment of high efficiency.

On the third of these sectors, the reduction of the use of primary energy for transport - and in particular the motor car - is technologically well within our capabilities by the use of more efficient transporters. But again it is an economic and social problem - will car drivers accept the smaller, lighter, less powerful cars with lower fuel consumption, unless taxation measures force them to do so? And will the UK Government move towards increased use of energy-efficient public transport? Unfortunately and depressingly, we think not.

11.5 OPTIONS FOR THE FUTURE

So in the light of these various influences and possible, if not probable, technical developments, what should our energy policy be? We first discuss generally the various strands of possible policy options.

11.5.1 Coal

It may well be that coal will be called upon to play a more important role in the United Kingdom and the European Community than at present, as it does in the rest of the world. Indeed, there are good reasons for reducing reliance on oil and gas, particularly when gas imports into Europe rise after the year 2000. Oil and gas are both imported from regions where political instability is possible and could lead to supply disruption and consequent price volatility. Clean coal technology has been steadily developed in the United Kingdom since 1960 but the initiative, at least for the time being, has moved to the European Community, with two demonstration plants nearing completion, one in Holland at Buggenum and one in Spain at Puertollano (with electricity outputs of between 200 and 300 MW_e, using coal gasification followed by clean-up, and then combined cycle generation). Generation efficiencies of around 45% are promised which compare favourably with the 35% efficiency of a conventional pulverized coal burning station with fuel gas desulphurization. This improved efficiency is essential if there is to be any hope of containing carbon dioxide emissions. Coal is the worst of the fossil fuels in terms of CO_2 production per unit of electricity delivered, twice as bad as natural gas. The prospects for absorbing CO_2 emissions from power stations are daunting; solvent extraction techniques or deep sea disposal in solid form could possibly double electricity generation costs.

Another fuel which may well play an increasingly important role is 'orimulsion', which is a water emulsion with heavy crude oil from Venezuela (the Orinoco Basin). Reserves are about 16% of world proved coal reserves. This fuel can also be gasified and used cleanly, as with clean coal technology. It might be wise if those constructing new gas-fired CCGT stations ensured there was enough space around the new stations to construct gasification plants and terminals to bring in coal or orimulsion!

The carbon dioxide problem could well dominate electricity generation over the next decade as planning decisions are taken for new generation plant. CHP systems are probably the most energy effective of all generation systems involving fossil fuels and consequently to be preferred, probably subsidized in some way, in order to reduce 'greenhouse' emissions of CO_2 world-wide (measured in kg/kWh of energy use, i.e. useful heat plus electricity).

11.5.2 Renewable and nuclear

It is clear, however, that stabilizing CO_2 emissions at 1990 levels by 2000 is going to be extremely difficult. In the United Kingdom it means cutting back by some 10 million tonnes of carbon per year over the next five years. Renewable energy and nuclear power have an important role to play but hydropower provides only 1% of UK electricity and is unlikely to increase substantially. The latest government target is 1500 MW of new renewables by 2000; that is, wind, landfill gas, waste incineration. In 1993 new renewables provided less than 0.5% of UK electricity but the potential is thought to be much higher, perhaps 10–20% by 2025. World Energy Council projections suggest the lower level. The market for renewables must be developed, and to this end renewable energy is heavily subsidized at present. It should continue to be so, well into the next century, a sensible strand of energy policy which has little to do with a policy based on straightforward competition.

We would add a particular point on renewables – that the development in the United Kingdom of tidal barrage schemes, as a medium- to long-term investment, could lead to the supply of several thousand megawatts of cleanly generated energy.

Nuclear power world-wide produces 6% of the energy supply or 17% of world electricity (in the United Kingdom the latter figure is 23% of electric power). Without nuclear power, world-wide CO_2 emissions from electricity generation would increase by about 28%. A clear view on the stabilizing and beneficial effects of nuclear power, leading to a phased expansion of internationally licensed stations, should be an important strand of electricity supply in the medium to long term. This should lead logically to the continuing development, not the abandonment, of the Fast Reactor programme as an international project. It is in the authors' view unwise to choose to continue to support the untried and 'engineeringly' difficult Fusion programme and to give up the almost mature technology of the Fast Reactor.

11.5.3 Transport

Perhaps the least tractable sector of energy use is transport; it is also the most polluting (in terms of 'greenhouse' gas emissions). The insatiable demand for private motor transport and the freedom it gives the individual traveller illustrates the sociopolitical nature of problems in energy planning. In the United Kingdom alone the number of cars is expected to increase by one third by the year 2020; in some developing countries car ownership will double or even treble. A policy of extending road provision in the United Kingdom, by doubling up dual carriageways to accommodate extra vehicles, is a policy of despair. The problem is essentially a car problem, although some goods would be better transported by rail than road. An efficient, flexible and imaginative public transport system will have to be developed; there are excellent role models such as the Metro system built on Tyneside some 10 years ago.

11.5.4 Technical research and development

As to the research required, we have described some of the technical developments which we feel are within our grasp as engineers. We would pick out the reduction of CO_2 production from coal plants as being the major technical problem to be solved, because of our conviction that coal must come back in the United Kingdom in due course. A solution may lie in better energy utilization through CHP; otherwise we must hope for some major world research and development in absorbing the CO_2 inevitably produced (the prospect of unrestrained large-scale coal generation in the developing world is a daunting one).

11.5.5 Non-technical problems

Engineers are good at solving technical problems; unfortunately many of the problems of energy use are dominated by social attitude based on ever-increasing material lifestyles which require ever-increasing amounts of energy and even encourage conspicuous waste. Much greater effort is required to solve these non-technical problems.

11.6 CONCLUSION

Medium- to long-term programmes require planning and political will; an energy policy that is predominantly market led will not accommodate them. The long lead times, typically 10 years, take them out of the influence of any market-led consideration.

However, it has to be recognized that the tremendous changes that have taken place already in electricity generation mean that as far as much of fuel utilization in the United Kingdom is concerned, the die is already cast to some extent - say the next 25–30 years. We are now committed to a virtual doubling of natural gas consumption for that period and a closing of many coal mines. The existing nuclear plants will continue to run as long as their available costs are low.

But more widely and in the longer term, we emphasize that future energy policy must be as much influenced by environmental factors, through legislation and regulation, as by technical developments.

There seems little doubt that, if we are to have a secure supply of energy into the next century which is reasonably priced, reliable and as environmentally friendly as we can make it, a balanced energy supply portfolio is absolutely necessary. It is also important to use our energy very much more efficiently than we do at the present time. Exhortation to use energy efficiently, even backed up by an information and training programme, is not enough; fiscal incentives are necessary. In the long run a change in social attitudes towards conspicuous waste of energy and other commodities will be necessary. (The changed attitude towards smoking over the past 10 years gives some hope of success.)

A simplistic market-led strategy does nothing for security of supply, the environment or energy-efficient use; quite the contrary in the latter case - the suppliers of electricity and gas make their profits by selling as much as possible. Regulation or legislation is already in place but this is an *ad hoc* approach to energy policy - the so-called light touch on the tiller recently advocated by the Government really will not do and is likely to be unacceptable in Europe where the long-term aim is to have an EU energy policy. That is not to say that competition and the market place have no place in energy policy; they are profoundly important instruments but they should be used to stimulate a move towards a strategic energy plan.

Within our recommended, environmentally conscious planned energy policy, we would advocate the following.

1. For electricity generation a fuel mix could be specified and the market encouraged to operate within that framework. There is already an obligation to provide 20% of UK electricity from non-fossil fuel sources, a mix of nuclear and renewables; subsidies are provided to encourage this end of the market, and this should be continued into the 21st century.
2. A suitable fuel mix would have a non-fossil fuel element (as now, including both nuclear and renewables for environmental and security reasons) of one third, one half evenly split between gas and coal, with the remaining one sixth as a 'swing' element of various fuels to balance the equation. (This way a company proposing to build a new gas-fired or coal-fired power station would bid for the licence; competition would thus be retained within the specified fraction of the system).
3. The policy should contain a strong energy efficiency element.
4. As an absolute necessity a transport policy should form an integral part of the overall strategy. This should minimize pollution and reduce, or at least stabilize, the use of cars through provision of more efficient public transport.

Index

Page numbers appearing in **bold** refer to figures and page numbers appearing in *italic* refer to tables.